Environmental Fate and Transport Analysis
with
Compartment Modeling

T0225524

Environmental Fate and Transport Analysis with Compartment Modeling

KEITH W. LITTLE

CRC Press
Taylor & Francis Group
Boca Raton London New York

CRC Press is an imprint of the
Taylor & Francis Group, an **informa** business

CRC Press
Taylor & Francis Group
6000 Broken Sound Parkway NW, Suite 300
Boca Raton, FL 33487-2742

First issued in paperback 2017

© 2012 by Taylor & Francis Group, LLC
CRC Press is an imprint of Taylor & Francis Group, an Informa business

No claim to original U.S. Government works

Version Date: 20120424

ISBN 13: 978-1-4398-8796-7 (hbk)
ISBN 13: 978-1-138-07413-2(pbk)

Library of Congress Cataloging-in-Publication Data

Little, Keith W.
  Environmental fate and transport analysis with compartment modeling / Keith W. Little.
    p. cm.
  Summary: "This book examines mathematical modeling and computer simulations that estimate the distribution of chemical contaminants in environmental media in time and space. Discussing various modeling issues in a single volume, this text provides an introduction to a specific numerical modeling technique called the compartment approach and offers a practical user's guide to the GEM. It includes the Generic Environmental Model (GEM) software package, which implements the techniques described. The author presents algorithms for solving linear and nonlinear systems of algebraic equations as well as systems of linear and nonlinear partial differential equations"-- Provided by publisher.
  Includes bibliographical references and index.
  ISBN 978-1-4398-8796-7 (hardback)
  1. Pollution--Mathematical models. 2. Transport theory--Mathematical models. 3. Diffusion--Mathematical models. 4. Cross-media pollution. 5. Compartmental analysis (Biology) 6. Pollutants. 7. Differential equations. I. Title.

TD174.L58 2012
363.7301'51--dc23   2012014569

Visit the Taylor & Francis Web site at
http://www.taylorandfrancis.com

and the CRC Press Web site at
http://www.crcpress.com

This book is dedicated to my father, Keith K. Little, who inspired me to be an engineer, and to my doctoral advisor and mentor, Donald T. Lauria, who gave me a passion for quantitative methods.

Contents

Preface

This book and the available software explain how to construct the equations that describe chemical fate and transport in environmental media (soil, water, air) in space and time and solve them using approximate numerical techniques. Readers can download the Generic Environmental Model (GEM) software and sets of input files for several of the presented examples at http://www.crcpress.com/product/isbn/9781439887967. The book and software have their origins nearly 30 years ago when I performed research related to parameter estimation in numerical surface-water quality models. The research was not concerned with the numerical modeling techniques per se but I needed computer code that would enable numerical simulations and could be integrated into a larger framework—optimization algorithms. Then, as now, it was very difficult to find straightforward, rigorous, and flexible software that would meet these needs, so I developed my own. I quickly came to appreciate the enhanced understanding of the scientific and mathematical issues to be gained when a modeler does not rely on a "black box" computer model and is forced to write his or her own mathematical equations and then solve them.

One reason that my early software was relatively easy to develop was that it was based on a numerical technique that had recently been introduced into the surface-water quality modeling field by Dr. Robert V. Thomann of Manhattan College, one of the fathers of modern water quality modeling and analysis. Thomann referred to his method as the "finite section" approach. It was rigorous, powerful, flexible, and, most important to me, intuitive and easy to understand. It seemed almost magical that one could simply hook the finite sections together, like Tinkertoys and build a sophisticated, numerical model of a complex environmental system. The finite section method has since been renamed by others as "control volume," "finite volume" (not to be confused with finite element), "box modeling," and "compartment modeling." Compartment modeling is the description used in this book.

Over the ensuing years, I found new and different applications for my compartment modeling program as my interests expanded beyond surface waters into multi-media modeling and the program evolved accordingly. I also gained an increased appreciation for the flexibility of the compartment approach in these new applications.

As one example of this flexibility, a recent application of the software by RTI International addresses chemical fate and transport of toxic constituents that are present in sewage biosolids along with beneficial nutrients. These biosolids are applied in agricultural settings to take advantage of the available nutrients, but the accompanying toxics may pose human health risks. The conceptual model involves a vertical soil column comprised of a set of compartments representing a farm field in which the topmost compartment periodically receives biosolids applications. An adjacent vertical soil column represents a downslope "buffer" area between the field and nearby surface water bodies. During the development of this compartment model, it became desirable to add the capability to simulate the interception of vertical mass transport caused by shallow engineered underdrains in the field. The solution?

Simply add another compartment representing the drainage system to accommodate this desired new feature.

Indeed, the compartment approach seemed a near-perfect platform for environmental system modeling in general, and I adopted a long-term goal of trying to synthesize and document the relevant theory scattered throughout the applied mathematics literature as the capabilities of my software expanded. Although the compartment method as applied to environmental sciences was discussed in several textbooks, those presentations were necessarily abbreviated because the textbooks were concerned with overall modeling techniques and environmental science—not with a specific numerical technique and certainly not from a generic multi-media environmental modeling view.

For example, the underlying differential equations that describe contaminant fate and transport in surface water systems and those for groundwater or soil systems are not very different; yet most textbooks and simulation software focus on one or the other. The compartment approach and solution techniques apply to both situations and I wanted a more integrated treatment. In addition, the technical jargon describing the underlying theory for numerically solving systems of partial differential equations, especially nonlinear systems, can quickly discourage non-mathematicians. Therefore, I also wanted a treatment of these techniques that was more user-friendly for environmental scientists and engineers.

My motivation, then, in writing this book was to summarize in one place the necessary mathematical principles and numerical techniques that enable straightforward, sophisticated, and flexible compartment modeling of environmental systems and make these techniques accessible to non-mathematicians. My aim in providing the software is to encourage both beginning and experienced modelers to go beyond "black box" approaches, that may shoe-horn a modeling problem into a pre-specified, often inappropriate, mold. I hope this book will allow them to take full ownership of the underlying equations and be able to solve them. It has been an adventure and a pleasure to write the book and develop the software, and it is my hope that others find them useful.

Keith W. Little
Raleigh, North Carolina

MATLAB® is a registered trademark of The MathWorks, Inc. For product information, please contact:

The MathWorks, Inc.
3 Apple Hill Drive
Natick, MA 01760-2098 USA
Tel: 508-647-7000
Fax: 508-647-7001
E-mail: info@mathworks.com
Web: www.mathworks.com

Mathematica® and the *Mathematica* logo are registered trademarks of Wolfram Research, Inc. (WRI–www.wolfram.com http://www.wolfram.com/) and are used herein with WRI's permission.

Acknowledgments

Many colleagues over the years have helped with the ideas and concepts in this book, and they are too numerous to list. Special thanks, however, go to Steve Chapra, Tufts University, Bob Ambrose, U.S. EPA (retired), and Michael Lowry and Steve Beaulieu, RTI International.

Author

Keith W. Little, PhD, PE, is a consulting engineer specializing in the development and application of mathematical modeling and systems analysis methods to environmental engineering and water resources problems. Dr. Little earned a PhD from the University of North Carolina at Chapel Hill, where, in 1985, he was awarded the Bernard Greenberg Award for Excellence in Doctoral Research. This research was the genesis of the GEM software used to illustrate the concepts in this book. Since then, the GEM has evolved in functionality and application and has been used to support risk assessment-based decision making at the U.S. Environmental Protection Agency (EPA).

Dr. Little has enjoyed a 30-year career in environmental engineering and research, including 15 years as a research environmental engineer at RTI International in Research Triangle Park, North Carolina, where he led the environmental modeling group and received the Science and Engineering Group's Annual Award for Exemplary Performance in 2005. He has authored numerous technical reports and articles for peer-reviewed journals. He has been active in various professional organizations and was president of the Colorado section of the American Water Resources Association in 1995. He is currently an independent consultant in Raleigh, North Carolina. Dr. Little may be contacted at keithwlittle@bellsouth.net.

1 Introduction

This book is about mathematical modeling and computer simulation to estimate the distribution of chemical contaminants in environmental media in time and space. Included is the Generic Environmental Model (GEM) software package that implements the techniques described.

This book is *not* an introductory text on environmental fate and transport modeling. There are many excellent references on that general topic elsewhere (Thomann, 1972; Thomann and Mueller, 1987; Mackay, 1991; Schnoor, 1996; Chapra, 1997; Ramaswami et al., 2005; Dunnivant and Anders, 2006). This book is an introduction to a specific numerical modeling technique—the "compartment" approach—as well as a user's guide to the GEM.

Environmental fate and transport modeling is essentially concerned with solving differential equations. Numerical methods[*] are used to solve these differential equations when the underlying equations are too complex to have exact (analytical) solutions. Unfortunately, the underlying differential equations of interest to environmental modelers are typically sufficiently complex as to require numerical solutions. Even more unfortunately for prospective environmental modelers (environmental professionals or students who desire to perform modeling but are relatively inexperienced in applied mathematics and/or computer programming), the numerical approach is generally considered to be so technical as to be daunting to all but professional modelers (Dunnivant and Anders, 2006). Our focus in this book and the accompanying software is to make numerical modeling accessible to the prospective environmental modeler. Readers can download the Generic Environmental Model (GEM) software and sets of input files for several of the presented examples at http://www.crcpress.com/product/isbn/9781439887967.

To illustrate the typical numerical modeling process, consider the classical finite difference method that involves approximating the derivative terms in the underlying differential equations with finite difference approximations, and then solving the resulting algebraic equations. Therefore, one must (1) first discretize the equations, a task that can be especially challenging if chemical concentrations vary in multiple dimensions. The discretized, algebraic system of equations must (2) be properly formatted and uploaded to a computer program that (3) invokes an internal equation solver subroutine or some procedure to solve the equations. Overlying all of these activities are (4) issues of numerical errors, e.g., unintentional dispersions or oscillations that can severely compromise a simulated solution, that must be first understood by the modeler and then controlled for by the parameters of the discretization

---

[*] Numerical methods are mathematical techniques for solving a wide variety of problems. They involve only arithmetic operations, which makes them ideally suited for digital computers. For example, your computer does not know how to evaluate an integral or a derivative, but you can help it by reducing this exercise to a sequence of simple arithmetic operations by using numerical methods.

process. Little wonder that prospective modelers shy away from this inscrutable scenario and turn to professional modelers.

Our goal for this book and the accompanying software is to make this overall four-step process less intimidating to prospective modelers. To the extent that a certain mathematical toolbox or repertoire is required for a proper understanding of numerical modeling, the modeler has no alternative but to attain this knowledge. Arguably, much worse than needing to resort to a professional modeler, is modeling incorrectly. A premise of this book is that there is nothing inherent in this mathematical repertoire that excludes those with only modest mathematical skills. We introduce these topics from the ground up, hopefully jargon-free (or jargon-explained), and require only some dedication on the part of the prospective modeler and very modest mathematical background. For example, we assume that the reader understands the concept of a derivative and has some familiarity with matrix notation, but make few other assumptions.

Regarding step (1), the complexity of discretizing the underlying differential equations numerically, enter the compartment approach. An environmental compartment is a volume of an environmental medium within which it is assumed that system parameters are constant and chemical concentrations do not vary spatially. The spatial domain (extent of the physical system to be modeled) of interest to the environmental contamination problem is conceptualized as a set of such compartments, the total number reflecting the desired spatial resolution and/or criteria to avoid or minimize numerical errors. The important point is that the modeler is discretizing the spatial domain itself rather than the differential equations.

This may not seem much of a distinction (and mathematically it is not), but this simple distinction goes a long way in making the numerical modeling process more approachable for a prospective modeler. Indeed, unlike the finite difference method, we approximate one-, two-, or three-dimensional systems completely naturally, simply by how we hook together the compartments comprising the spatial domain. The underlying mathematics is indifferent to the spatial dimensionality and the configuration of the modeler's compartments implicitly determines the dimensionality of the solution. The compartment approach makes discretizing the spatial domain of the model completely natural and intuitive. What remains for dynamic problems, however, is to also discretize the temporal derivative. We use finite differences for that and those techniques are developed, explained, and illustrated from first principles in this book.

Steps (2) and (3) represent the "bookkeeping" required to set up a numerical model, format the equations, and invoke an equation solver. The GEM performs those tasks. Step (4), understanding the potential for numerical errors and properly controlling the parameters of the numerical solution, is the user's responsibility. Those issues are developed and explained from first principles in this book and illustrated with numerous examples using the GEM. Therefore, the user's responsibility is to understand the system, overall numerical modeling process and potential for errors, and provide the geometry and other parameters of the model in the proper GEM input files. The GEM handles the bookkeeping and solves the equations.

In summary, the GEM is essentially an equation solver that focuses on mass balance equations describing chemical fate and transport in environmental media. It allows the user to easily assemble the equations as desired to represent the medium or media of interest and the relevant fate and transport processes, and then solves those systems of algebraic (for steady-state problems) or differential equations (for time-variable or dynamic problems) using numerical methods.

In contrast, most other environmental fate and transport software packages assume a specific medium (medium-specific models) or a specific mix of media (multi-media models) and impose a relatively fixed set of fate and transport processes. Making even minor changes to these models can be very challenging. Indeed, the underlying system of equations to be solved is often invisible and largely unknown to the user. Thus, the fundamental difference in architecture offered by the GEM is that it allows much greater freedom in specifying the conceptual model and processes, is generic rather than specific to a particular medium, and never separates the modeler from the equations being solved. With this flexibility, it becomes relatively simple to build and solve very complex models in environmental settings of the user's choice.

This book can be considered an introduction to environmental compartment modeling, a user's guide to the GEM, or both. As a compartment modeling reference, the GEM is used extensively throughout to illustrate theory with numerous examples. As a user's guide, the theoretical discussions illuminate and document the GEM's functionality. The remainder of this introduction provides a further overview of the GEM's capabilities and illustrates them with two very different examples.

1.1 GEM HIGHLIGHTS

An overview of the major functionality and limitations of the GEM follows.

- The GEM can be used as an environmental modeling system or simply as an equation solver. In the *Environmental System* mode, the GEM builds a set of chemical mass balance equations that describe fate and transport in environmental media based on user input files and subsequently solves those equations. In *Equation Solver* mode, the user enters the equations of interest—that may have nothing to do with environmental modeling—and the GEM subsequently solves those equations.
- In both modes, the set of equations can be algebraic or differential, linear or nonlinear.
- The GEM code does not impose limits on the number of equations that can be solved.
- For linear, algebraic equations, standard linear algebra algorithms are used.
- For nonlinear, algebraic equations, a robust quasi-Newton algorithm is included.
- For solving systems of differential equations, four different solution options are included (three for the *Equation Solver* mode) along with an extensive review of their advantages and disadvantages and error characteristics. The quasi-Newton method is incorporated into these methods as appropriate for nonlinear systems.
- The GEM is available in the Visual Basic 6.0 (VB 6) programming language and is included on this book's website at www.crcpress.com/product/isbn/9781439887967 as an executable (machine language) file. Users wanting a

copy of the source code should contact the author to discuss cost and conditions of use for their individual applications. A Java (Sun Microsystems) version is under development at this writing. Check book's website for updates.
- The GEM reads input data from and writes data to comma-separated value (.csv), flat files such as produced by Microsoft Excel or may be generated using a text editor. Thus, the program is independent of links to proprietary database software and is highly flexible for inputs and outputs.
- For dynamic problems, any of the GEM inputs and parameters in the input files may be updated "on the fly" at each time step using a shell procedure that requires the user to write and compile external software to perform the updating. For dynamic loadings, the shell procedure can be used or updates can be read from a *loads.csv* file without requiring an external program.

The following apply to use of the GEM in *Environmental System* mode.

- The GEM evolved from surface water fate and transport applications, but has application throughout environmental media. The medium can be water, air, a porous medium such as a groundwater system, or combinations of media (multi-media modeling).
- The GEM is not a hydrodynamic or hydraulic model. Only chemical mass balance equations are solved. Flow regimes and dispersion/diffusion coefficients must be provided by the user in input files and/or updated externally during run time using the shell procedure.
- The GEM assumes that the transport processes (advective flow and dispersion/diffusion fluxes) are linear with respect to chemical concentrations. Dispersive/diffusive fluxes between compartments are assumed to be driven by concentration gradients in accordance with Fick's law. The source and sink terms can be linear or nonlinear.
- A single chemical or a system of multiple, interacting chemicals can be modeled. For multiple chemicals, the user may specify which chemicals are relevant to which compartments.
- Each compartment is spatially uniform with respect to concentrations and parameters. Concentration gradients are allowed only among compartments, not within a compartment. The numbers and sizes of compartments are user-specified.
- The spatial dimensionality is arbitrary. The way the compartments are arranged, (zero-, one-, two-, or three-dimensional) is specified by the user.
- Three types of boundary conditions are included in the GEM: (1) fixed concentration, (2) zero gradient, and (3) linear gradient.

In *Equation Solver* mode, the GEM is similar in principle to other generic mathematics-oriented software packages such as *Mathematica*®, MathCad, MATLAB®, and Stella. There are undoubtedly advantages and disadvantages in using the GEM in *Equation Solver* mode versus a commercial package. For example, Stella appears to be limited to a single type of dynamic numerical solution (comparable to our forward time [Euler] method discussed in Chapter 6) while the GEM offers several options. However, the real advantage of the GEM is its *Environmental System* mode. In this mode, by automatically doing the extensive bookkeeping associated with environmental fate and transport equations, the GEM allows easy construction and solution of these equations.

1.2 GETTING STARTED: TWO EXAMPLES

We introduce the GEM with two examples to illustrate its capabilities.

1.2.1 EXAMPLE OF CHEMICAL RISK ASSESSMENT PROBLEM

The first example is a chemical risk assessment problem for a single toxic chemical in a multi-media environment. The example is based on actual risk assessment modeling undertaken by the U.S. Navy in 2005 (NEHC, 2006) to evaluate potential human health risks associated with sinking the *USS Oriskany* as an artificial reef. An overview of the vessel's history and the risk assessment setting follows.

The construction of the *USS Oriskany* was completed shortly after World War II at the New York Navy Yard. She was the third U.S. Navy ship to bear the Oriskany name, in tribute to a major battle during the U.S. revolutionary war. She was one of 24 Essex class aircraft carriers.

The *Oriskany* saw early service in the Korean War and continued to serve in the Pacific Fleet during the Vietnam conflict. She was decommissioned in 1976, sold for scrap in 1995, and subsequently repossessed by the Navy in 1997. In 2004, in response to an increase in demand for increased aquatic habitats in coastal areas, the decision was made to sink her as an artificial reef some 20 miles off Pensacola, Florida in the Gulf of Mexico.

Even after extensive removal of toxic materials prior to sinking, it was estimated that approximately 340 kg (750 lb) of polychlorinated biphenyls (PCBs) would remain on board, primarily in solid form contained in the insulation layers of electrical cable throughout the ship. The presence of these toxic materials and the potential for their leaching into the water column and contamination of the food web prompted the U.S. EPA to require a risk-based PCB disposal permit.

The risk assessment was based on the predicted water column concentrations after leaching began. The assessment was developed from the Prospective Risk Assessment Model (PRAM), a computer program developed specifically for the *Oriskany* scenario and based on the "fugacity" approach (see Box 1.1). Completion and peer review of the PRAM and risk assessment resulted in the judgment that health risks were acceptable, and the *Oriskany* was sunk in May 2006. She remains the largest artificial reef to date, popularly known as the "Great Carrier Reef," and is one of the top diving sites in the world. Figures 1.1, 1.2, and 1.3 depict the *Oriskany* before, during, and after sinking.

**BOX 1.1   FUGACITY MODELING AND
GEM'S EQUATION SOLVER OPTION**

The PRAM was developed as a Level III fugacity model. Fugacity is a chemical-specific thermodynamic property that represents the partial pressure of a chemical in a medium. Fugacity capacity reflects the ability of the medium to absorb the chemical. The analogy is often made relating fugacity to heat transfer problems, in which a medium's fugacity is analogous to the temperature of a substance, and the medium's fugacity capacity is analogous to the heat capacity of the substance. At thermodynamic equilibrium between two media, their fugacities (temperatures) will be equal, but their fugacity capacities (heat capacities) will not. At equilibrium, just as at equal temperature for the heat transfer analogy, there is no tendency for one medium to lose chemical mass (heat) to another medium.

The fugacity approach for environmental modeling—especially multi-media modeling—was popularized by Donald Mackay (1991) who (1) summarized the fugacity capacities of various environmental media, (2) categorized the approach into four different levels of increasing complexity, and (3) presented the theory (and accompanying computer programs) in a tidy, holistic framework. Fugacity modeling is not a shortcut to multi-media (or any) environmental modeling. It requires the same information as a comparable concentration-based approach (Trapp and Matthies, 1998). However, thinking of various media as possessing different capacities to store chemicals is an insightful concept, and the Level I through IV categorization and summary of fugacity capacities provided by Mackay are found helpful by many modelers.

The GEM's *Environmental System* option is concentration-based and therefore does not support fugacity modeling. However, fugacity models can be developed offline and then solved in the GEM using the *Equation Solver* option.

Extensive development went into the PRAM—not because new science was involved, but rather due to the lack of an existing model that would adequately capture the environmental setting and support the desired risk assessment. The GEM is designed for just such a situation and our first getting-started example is to build a PRAM-like model using the GEM.[*]

The PRAM considered the artificial reef environment as a set of five compartments (Figure 1.4) comprised of (1) the submerged ship compartment containing the source[†] of leaching polychlorinated biphenyls (PCBs), (2) a lower water column compartment surrounding the ship compartment and extending from the ocean bed to the pycnocline (a horizontal layer effectively separating upper, warmer waters from lower, colder, denser waters), (3) a shallow (0.1 m) sediment compartment lying beneath the

---

[*] The author served as a technical peer reviewer for the development of the PRAM and the associated Time Dynamic Model (TDM). The GEM was used for benchmarking these models.
[†] Emission rates of PCB from various sources within the ship were estimated by laboratory studies using representative on-board materials.

FIGURE 1.1 *USS Oriskany.*

FIGURE 1.2 *Oriskany* going down.

footprint of the upper water compartment, (4) an upper water column compartment with the same footprint as the lower water compartment and extending from the pycnocline to the water surface, and (5) an air compartment with the same footprint. Each compartment in the PRAM is assumed to be completely mixed (no PCB concentration gradients exist within a compartment; they exist only among compartments).

Within these five compartments, the PRAM considered PCBs in three phases: (1) dissolved in the water and air, (2) sorbed to particulates, and (3) sorbed to

FIGURE 1.3   Portion of island structure after sinking.

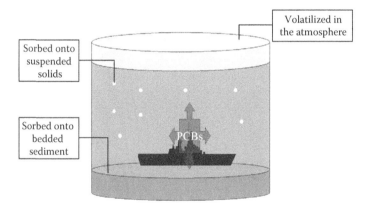

FIGURE 1.4   Compartments for abiotic PRAM. (From U.S. Navy Environmental Health Center. 2000. *Preliminary User's Manual: Draft Prospective Risk Assessment Model (PRAM): Version 2.0.* With permission.)

dissolved organic carbon. Particulates in the water are suspended solids; they are aerosols in the air and bed solids in the sediment compartment.

In addition to the abiotic, five-compartment model described above, the PRAM also simulated the uptake of PCBs throughout the food chain expected to develop on the artificial reef. Estimated concentrations in higher trophic-level fish then provided the exposure levels for the recreational fisher receptor for the risk assessment.

We simplify these assumptions for the GEM example. We have only four completely mixed compartments by assuming only a single water column compartment extending from the ocean floor to the surface. We assume that no particulates or dissolved organic material are present in the ship, water column, or air compartments;

therefore no sorbed PCBs are present there. The sediment compartment contains solids, of course. We are interested only in the abiotic portion of the PRAM model.

A mass balance of PCBs within each compartment results in a set of four simultaneous, linear equations. If we assume steady-state conditions (no changes over time), these are algebraic equations with single solutions (PCB concentration in each compartment). If we assume these concentrations can vary over time (because of a variable emission rate or other temporal variation), the four equations are differential, yielding a time series of compartment-specific concentrations upon solution. Let's develop these equations.

A steady-state, mass balance of PCB in the ship compartment (compartment 1) results from setting PCB inputs equal to outputs. The mass input is the PCB[*] leach rate from the various PCB-laden materials remaining on board. Let this leach rate be denoted as W with units of mass per time (M/T). The outputs are the losses of PCB mass from the compartment due to inter-compartment transport processes and PCB biodegradation. For the ship compartment, we assume that the only inter-compartment transport process is dispersion/diffusion.[†] While there is an ocean current (advective transport), we assume that it is not flowing through the ship to any significant degree. Dispersion/diffusion mass transport occurs in the direction of a concentration gradient and is expressed mathematically as a flow rate times the difference in concentration between two compartments. At steady state, the inputs must balance the outputs, or

$$W + \frac{E_{1,2}A_{1,2}}{L_{1,2}}(c_2 - c_1) - kV_1c_1 = 0 \tag{1.1}$$

where

c_1 = PCB concentration in ship compartment
c_2 = PCB concentration in water column compartment
$E_{1,2}$ = dispersion/diffusion coefficient between ship compartment (1) and water column compartment (2) (L^2/T)
$A_{1,2}$ = surface area between compartments 1 and 2 (L^2)
$L_{1,2}$ = distance across which dispersion/diffusion is assumed to occur (L)
k = first order biodegradation rate constant (T^{-1})
V_1 = volume of ship compartment (L^3)

Note for the dispersion/diffusion term that if the concentration gradient is negative, i.e., c_1 is greater than c_2 (which it will be because compartment 1 is the PCB source), the dispersion/diffusion term will have a negative sign and represent a loss from the

---

[*] There is no single PCB chemical. PCBs represent a set of homologs or compounds that have similar properties but somewhat different molecular structures. The PRAM analysis was specific to each homolog. We use the PCB term as an arbitrary example homolog.

[†] Diffusion and dispersion are different processes. Where diffusion acts at the molecular level, dispersion is a macro process caused by non-uniform flow patterns and as a mass transfer process would typically dwarf molecular diffusion. Nonetheless, they are modeled identically and we will simply denote this collective mass transport term as dispersion/diffusion. The diffusion/dispersion from the ship compartment/water column and water column/sediment interfaces is primarily diffusion due to the absence of significant flow while across the water column/air interface dispersion would predominate.

compartment, not a gain. Note also that the units in each term of Equation (1.1) are consistent, i.e., M/T.

If we assume that these processes are not at steady state, the inputs minus outputs in Equation (1.1) do not equal zero, but rather equal a derivative term representing the time rate of PCB mass accumulation within the compartment, or

$$V_1 \frac{dc_1}{dt} = W + \frac{E_{1,2}A_{1,2}}{L_{1,2}}(c_2 - c_1) - kV_1c_1 \tag{1.2}$$

where

$$\frac{dc_1}{dt} = \text{time derivative } (M/(L^3 - T)).$$

Now consider the single water column compartment (compartment 2) where we assume an advective transport mechanism due to the prevailing ocean current and dispersion/diffusion transport. The source of PCBs to the water column is the dispersion/diffusion from the ship compartment through the exterior openings in the hull and superstructure (an area far smaller than the total surface area of compartment 1). All other processes are losses from the water column to adjoining compartments. The steady-state mass balance equation is then

$$\frac{E_{1,2}A_{1,2}}{L_{1,2}}(c_1 - c_2) + \frac{E_{2,3}A_{2,3}}{L_{2,3}}(F_dc_3 - c_2) + \frac{E_{2,4}A_{2,4}}{L_{2,4}}(c_4 - c_2) - Q_2c_2 - kV_2c_2 = 0 \tag{1.3}$$

where Q_2 = prevailing ocean current advective flow rate (L^3/T).

Note that in the dispersion/diffusion term between compartments 1 and 2 seen above in Equation (1.2), we reversed the concentration gradient. If c_1 is greater than c_2 (which, again, it will be), this term is positive and represents an input to compartment 2. We also have two additional dispersion/diffusion terms—between the water column and the sediment compartment (compartment 3), and between the water column and the overlying air compartment (compartment 4). The corresponding dynamic (i.e., differential) mass balance equation for the water column compartment is

$$V_2 \frac{dc_2}{dt} = \frac{E_{1,2}A_{1,2}}{L_{1,2}}(c_1 - c_2) + \frac{E_{2,3}A_{2,3}}{L_{2,3}}(F_dc_3 - c_2) + \frac{E_{2,4}A_{2,4}}{L_{2,4}}(c_4 - c_2) - Q_2c_2 - kV_2c_2$$
$$\tag{1.4}$$

Similarly, for the sediment compartment (compartment 3), the source of PCBs is dispersion/diffusion from the water column and all other processes are losses (outputs). As in the ship compartment, there is no advective flow through the sediment. The steady-state mass balance is

$$\frac{E_{2,3}A_{2,3}}{L_{2,3}}(c_2 - F_dc_3) - kV_3c_3 = 0 \tag{1.5}$$

Because the sediment bed contains solids and PCBs readily sorb to solids, we are now distinguishing between dissolved and sorbed PCB concentrations in the sediment layer. c_3 represents the total (dissolved + sorbed) concentration and F_d is the fraction of the total concentration dissolved. Only the dissolved fraction is available for dispersion/diffusion. Note in the biodegradation term that we have not included F_d; therefore we are implicitly assuming for further simplicity that both dissolved and sorbed PCBs undergo biodegradation and at the same rate k. The corresponding dynamic mass balance equation for the sediment compartment is

$$V_3 \frac{dc_3}{dt} = \frac{E_{2,3}A_{2,3}}{L_{2,3}}(c_2 - F_d c_3) - kV_3 c_3 \qquad (1.6)$$

Finally, for the air compartment (compartment 4), the PCB source is dispersion/diffusion from the water column (volatilization) and all other processes are losses. As with the water column, a loss term represents net advective flow, which in this case is the prevailing air current. We assume that biodegradation is not a significant process for PCBs in air. The steady-state mass balance equation is then

$$\frac{E_{2,4}A_{2,4}}{L_{2,4}}(c_2 - c_4) - Q_4 c_4 = 0 \qquad (1.7)$$

where Q_4 = prevailing air flow rate (L^3/T). The corresponding dynamic mass balance equation is

$$V_4 \frac{dc_4}{dt} = \frac{E_{2,4}A_{2,4}}{L_{2,4}}(c_2 - c_4) - Q_4 c_4 \qquad (1.8)$$

Figure 1.5 depicts how the four compartments were arranged for the GEM simulation. Compartments 5 and 6 are boundary and dummy compartments, respectively, and are described in Chapter 3.

In an actual modeling project, realistic parameter estimation is generally the most time-consuming activity. Parameter estimation involves calibration of the model to available data (if the modeled system exists and data are available) or best professional judgment based on well-founded science, sensitivity analyses, and uncertainty analyses. For our example, however, we simply adopt the following parameter values.[*]

V_1 = 120,000 m³ (300 m × 20 m × 20 m)
V_2 = 1,080,000 m³ (300 m × 60 m × 60)
V_3 = 1,800 m³ (300 m × 60 m × 0.1 m)
V_4 = 1,080,000 m³ (300 m × 60 m × 60 m)

---

[*] We make no particular effort to echo the PRAM's parameter values. Again, our GEM example is not a reproduction of the PRAM, but rather an illustration of how similar functionality is easily incorporated into the GEM framework. Modeled results from this example are simply for illustration and do not reproduce previous results from the PRAM effort.

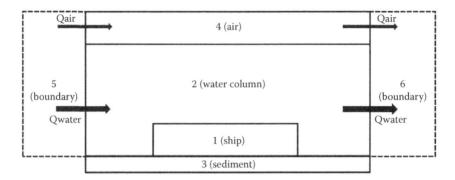

FIGURE 1.5   Compartments for PRAM-like model.

$A_{1,2} = 1,800$ m^2 (three sides with 10% openings in hull)
$A_{2,3} = 12,000$ m^2 (excluding ship footprint)
$A_{2,4} = 18,000$ m^2
$L_{1,2} = 1$ m
$L_{2,3} = 0.1$ m
$L_{2,4} = 10$ m
$E_{1,2} = 10$ m^2/day
$E_{2,3} = 1$ m^2/day
$E_{2,4} = 100$ m^2/day
$Q_2 = 10,000$ m^3/day
$Q_4 = 100,000$ m^3/day
$k\quad = 0.01$/day
$F_d\quad = 0.1$
W (steady state) = 1,000 g/d

With these parameter values, we can write the steady-state mass balance equations as the following set of four simultaneous linear algebraic equations. The four concentrations are the unknowns to be determined by the mathematical solution.

$$1,000 + 18,000(c_2 - c_1) - 1,200c_1 = 0 \tag{1.9a}$$

$$18,000(c_1 - c_2) + 120,000(0.1c_3 - c_2) +$$
$$180,000(c_4 - c_2) - 10,000c_2 - 10,800c_2 = 0 \tag{1.9b}$$

$$120,000(c_2 - 0.1c_3) - 18c_3 = 0 \tag{1.9c}$$

$$180,000(c_2 - c_4) - 100,000c_4 = 0 \tag{1.9d}$$

For the dynamic problem, we add the temporal derivative term to each of the above equations. For compartment 1 where the PCB loading is, we denote the dynamic loading as $W(t)$, which we will quantify later. The dynamic problem then consists of four simultaneous differential equations. The four concentrations remain the unknowns but rather than a single time-invariant solution as for the steady-state case, the solution will yield a time series of concentrations in each compartment.

$$120,000\frac{dc_1(t)}{dt} = W(t) + 18,000(c_2 - c_1) - 1,200c_1 \tag{1.10a}$$

$$1,080,000\frac{dc_2(t)}{dt} = 18,000(c_1 - c_2) + 120,000(0.1c_3 - c_2) + \tag{1.10b}$$
$$180,000(c_4 - c_2) - 10,000c_2 - 10,800c_2$$

$$1,800\frac{dc_3(t)}{dt} = 120,000(c_2 - 0.1c_3) - 18c_3 \tag{1.10c}$$

$$1,080,000\frac{dc_4(t)}{dt} = 180,000(c_2 - c_4) - 100,000c_4 \tag{1.10d}$$

After the structural form of the equations has been defined and the various parameters have been estimated, the "science" is done and the numerical methods needed to solve the equations come into play. Here also is where a prospective modeler often encounters trouble. Solving a system of linear simultaneous algebraic equations (steady-state system) or advancing this solution forward in time (even more difficult dynamic solution) presents obvious difficulties[*] to the modeler who does not understand the underlying numerical techniques and/or the one who lacks access to the necessary computer code. The GEM is designed both to help with the construction of the underlying modeling equations (the science) and contains algorithms to effect their solutions for steady-state or dynamic problems.

To continue the example, using the GEM's internal linear algebraic equation solver, the solution to the steady-state system of equations is

c_1 (ship) = 6.23×10^{-2} g/m^3
c_2 (water column) = 1.09×10^{-2} g/m^3
c_3 (sediment) = 1.08×10^{-1} g/m^3
c_4 (air) = 6.98×10^{-3} g/m^3

As you might expect, among the ship, water column, and air compartments, the ship compartment has the highest concentration because it is the PCB source. The water column concentration is approximately one-half an order of magnitude less and the air compartment concentration is essentially an order of magnitude less. The total

---

[*] Solving four simultaneous linear algebraic equations may not seem particularly difficult. The substitution method will certainly work. However, what if 40 or 40,000 equations were involved? Those numbers are not unrealistic for many numerical models. Substitution quickly becomes impossible under those conditions.

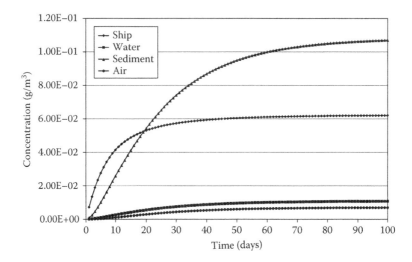

FIGURE 1.6   Time to steady state.

sediment compartment concentration (dissolved plus sorbed), however is nearly twice that of the ship compartment, reflecting the trapping aspect of the sediments because 90% (F_d = 0.1) of the sediment concentration is tied up in sorption to the sediments.

To illustrate a dynamic solution, we again assumed the PCB loading W was a constant at 1,000 g/day and ran the dynamic model to see how long the four compartments took to reach steady state, starting from initial concentrations of 0. Using the BTBS method (see Chapter 6) and a 1-day time step, the time series of concentrations in the four compartments are shown in Figure 1.6. The ship compartment reaches steady state relatively quickly, after 30 to 40 days. It takes considerably longer for the other compartments to reach steady state. This lag reflects the bottleneck represented by the relatively slow diffusive/dispersive transport through the openings in the ship. The slower mass transport between the ship and the other compartments, the longer it will take for the other compartments to reach steady state.

We ran a second dynamic solution that more realistically assumes that the PCB emission rate will slowly decline over time as the PCB materials gradually lose their load. We assumed that this occurs as a first-order process, with a decay constant of 0.01/day, and the initial emission rate is again 1,000 g/day. Thus, for this loading scenario, $W(t) = 1,000e^{-0.01t}$. We again ran the GEM using a 1-day time step and the resulting concentration time series are shown in Figure 1.7. What we see now is that the ship compartment concentration reaches a maximum—somewhat less than the previous steady-state value—after some 40 days and then decreases in accord with the decaying PCB emission rate. The other compartments follow suit, but lag the ship compartment somewhat. This lagging pattern again reflects the transport bottlenecks from the ship to the rest of the system. If it takes longer for these other compartments to reach their maximum concentrations because of the bottleneck, it will similarly take them longer to depurate their PCB loads.

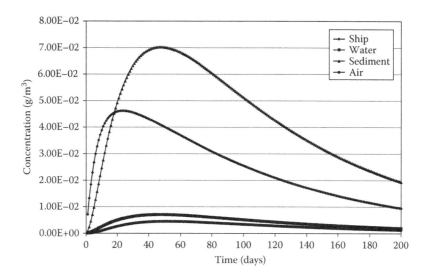

FIGURE 1.7   Results with decaying source.

1.2.2   EXAMPLE OF MATHEMATICAL PROBLEM

Our second introductory example is a purely mathematical problem—it has nothing to do with environmental modeling. The implacable use of "purely mathematical examples" by professors in math classes turns many otherwise good students off mathematics ("What is it good for?!"). Nonetheless, we use such an example here to illustrate the GEM's flexibility in *Equation Solver* mode. The problem is to find the solution to the following two nonlinear equations in two unknowns (Matthews, 1992):

$$x^2 - 2x - y + 0.5 = 0$$
$$x^2 + 4y^2 - 4 = 0$$

(1.11)

As described in later chapters, the GEM's steady-state system of nonlinear equations to be solved is given by a matrix[*] Equation (1.12) where, in *Environmental System* mode, the $\underline{c}$ vector is the set of unknown chemical concentrations, the **A** matrix contains the linear fate and transport parameters (e.g., flows, dispersion coefficients, first-order rate coefficients), the $\underline{f}(\underline{c})$ vector contains nonlinear source and sink terms, and the $\underline{b}$ vector is the forcing function of external loads. (For those not totally comfortable with matrix notation, we gently introduce those concepts in later chapters.)

$$\mathbf{A}\underline{c} + \underline{f}(\underline{c}) = -\underline{b}$$

(1.12)

---

[*] We introduce matrix notation gently as this book proceeds. In the interest of brevity for this introduction, we simply assert the equations.

Equation (1.12) is perfectly general, i.e., not limited to environmental modeling and we can express our example problem in the same format using the GEM's *Equation Solver* mode as

$$
\begin{bmatrix} -2 & -1 \\ 0 & 0 \end{bmatrix} \begin{bmatrix} x \\ y \end{bmatrix} + \begin{bmatrix} x^2 \\ x^2 + 4y^2 \end{bmatrix} = \begin{bmatrix} -0.5 \\ 4 \end{bmatrix}
\tag{1.13}
$$

and solve the nonlinear, steady-state problem (1.13) using the iterative quasi-Newton algorithm described in later chapters. Using a stopping criterion[*] of 0.0001 and a starting guess of $x = y = 10$, the GEM converged to $x = 1.90$ and $y = 0.311$ after seven Newton iterations, which is the correct solution.

1.3   ORGANIZATION AND SUGGESTED READING STRATEGY

The remainder of this book is organized as follows. Chapter 2 is the user's guide to the GEM. Chapter 3 summarizes the compartment approach to numerical environmental modeling and describes the development of the transport terms for the GEM and its boundary conditions. Chapter 4 describes source and sink terms. Chapter 5 presents solution techniques for steady-state problems and discusses error characteristics. Chapter 6 presents solution techniques for dynamic problems and additional error characteristics. Numerous examples are included throughout to demonstrate the methods.

Some readers may already be familiar with the theory underlying numerical models and the compartment approach and are interested only in using the GEM in their applications. Those readers might opt to read the Chapter 2 user's guide and consider Chapters 3 through 6 as reference material on an as-needed basis. Other readers may have no interest in running the GEM and are interested in the underlying numerical modeling theory of the compartment approach. Those readers might jump right to Chapter 3.

REFERENCES

Chapra, S.C. 1997. *Surface Water-Quality Modeling: Preliminary Edition.* New York: McGraw-Hill.
Dunnivant, F.M. and Anders, E. 2006. *A Basic Introduction to Pollutant Fate and Transport.* New York: Wiley-Interscience.
Mackay, D. 1991. *Multimedia Environmental Models: The Fugacity Approach.* Chelsea, MI: Lewis Publishers.
Matthews, J.H. 1992. *Numerical Methods for Mathematics, Science, and Engineering*, 2nd ed. Englewood Cliffs, NJ: Prentice Hall.
NEHC. 2000. *Preliminary User's Manual: Draft Prospective Risk Assessment Model (PRAM), Version 2.0.* Norfolk, VA: U.S. Navy Environmental Health Center.

---

[*] The system of equations is not actually solved in the format shown, but rather is expressed as an implicit function $\underline{\mathbf{g}}(\underline{c})$ where $\underline{\mathbf{g}}(\underline{c}) = \mathbf{A}\underline{c} + \underline{\mathbf{f}}(\underline{c}) + \underline{\mathbf{b}} = \underline{\mathbf{0}}$ and the roots of $\underline{\mathbf{g}}(\underline{c})$ are found. Thus, the stopping criterion tells the algorithm how close to 0 is acceptable. But, we get ahead of ourselves.

NEHC. 2006. *Ex-Oriskany Artificial Reef Project Prospective Risk Assessment Model (PRAM), Version 1.4c.* Final Report. San Diego: U.S. Navy Environmental Health Center.

Ramaswami, L., Anu, J.B., and Milfor, J.M. 2005. *Small, Integrated Environmental Modeling: Pollutant Transport, Fate, and Risk in the Environment.* New York: John Wiley & Sons.

Schnoor, J.L. 1996. *Environmental Modeling: Fate and Transport of Pollutants in Water, Air, and Soil.* New York: John Wiley & Sons.

Thomann, R.V. 1972. *Systems Analysis and Water Quality Management,* New York: McGraw-Hill.

Thomann, R.V. and Mueller, J.A. 1987. *Principles of Surface Water Quality Modeling and Control.* New York: Harper & Row.

Trapp and Matthies. 1998. *Chemodynamics and Environmental Modeling: An Introduction.* Berlin: Springer.

2 GEM User's Guide

This chapter is the user's guide to the GEM. Successful use of the GEM requires understanding how the input files are populated, and confident use further requires understanding the mathematical foundations of the model and solution techniques and potential errors that can result. How the files are populated is the objective of this chapter. The mathematical foundations and error characteristics are discussed in Chapters 3 through 6.

Reading software user's guides is about as exciting as reading the latest IRS tax code changes, and this one may be no different. However, we attempt to enliven the presentation by continuing the getting-started PRAM-like example from Chapter 1. The example is an interesting environmental problem, and is sufficiently compact that its input files are readily presented. Before beginning the user's guide, however, we embellish the PRAM-like example by formulating it in two other ways that are alternatives to the Chapter 1 F_d-based approach.

It may seem strange to begin a user's guide with yet more examples, but our purpose is threefold. First, the three alternative formulations collectively provide a diverse pool of examples from which much of the GEM's capability can be illustrated as we describe the various input files. Second, there is a topic—the distinction between steady-state and equilibrium conditions—that needs clarification before we go much further in this book. Our second alternative PRAM-like model formulation (k_s/k_{ds} version) will illustrate these differences. Finally, apart from GEM documentation objectives, these three alternatives are illustrative of the variety of modeling constructs encountered in environmental modeling.

2.1 TWO ALTERNATIVE FORMULATIONS OF PRAM-LIKE MODEL

2.1.1 k_s/k_{ds} ALTERNATIVE

Most prospective modelers understand the difference between steady-state and dynamic problems. However, widespread misunderstanding—or at least ambiguity—exists between steady-state and equilibrium conditions. Steady-state and equilibrium conditions are not the same. Steady state indicates no changes with respect to time. Equilibrium means that two interacting chemicals (e.g., dissolved and particulate) show no net change over time. Aren't these equivalent statements? No. A system can be unchanging in time (steady-state) but some "bottleneck" (a mass transport bottleneck or a kinetic bottleneck) prevents sufficient contact time between the two interacting chemicals to allow them to attain their equilibrium concentrations. A system can be at steady state and equilibrium, at steady state but not at equilibrium, at equilibrium but not at steady state, or, indeed, at equilibrium and changing over time.

To aid in illustrating these points, our second alternative PRAM-like model considers the sorbed and dissolved components in the sediment compartment *explicitly* and describes their interaction through first-order kinetics that represent the sorption and desorption processes. Recall Equation (1.6), which is the mass balance equation for total chemical in the sediment compartment:

$$V_3 \frac{dc_3}{dt} = \frac{E_{2,3}A_{2,3}}{L_{2,3}}(c_2 - F_d c_3) - kV_3 c_3 \qquad \text{(1.6 repeated)}$$

We now replace this equation by two equations, one for dissolved chemical and one for sorbed chemical including new source and sink terms:

$$V_3 \frac{dc_{d3}}{dt} = \frac{E_{2,3}A_{2,3}}{L_{2,3}}(c_2 - c_{d3}) - kV_3 c_{d3} + k_{ds}V_3 c_{p3} - k_s V_3 c_{d3}$$

$$V_3 \frac{dc_{p3}}{dt} = k_s V_3 c_{d3} - k_{ds}V_3 c_{p3} - kV_3 c_{p3}$$

where
c_{d3} = dissolved chemical (Mc/L$^3_{total}$)
c_{p3} = particulate (sorbed) chemical (Mc/L$^3_{total}$)
k_{ds} = first-order desorption rate constant[*] (/T)
k_s = first–order sorption rate constant (/T)

Note that the second new equation has no diffusion term because we are assuming (as in the original example) that sorbed chemical does not diffuse. Also note that we continue to assume that all the chemical in the sediment compartment (dissolved and sorbed) is subject to the same first-order biodegradation loss mechanism with rate constant k.

Finally, note that we are expressing both the dissolved and sorbed concentrations in the same units,[†] i.e., chemical mass per total volume (Mc/L^3_T). It is common in practice to express the dissolved concentrations as chemical mass per volume of water (Mc/L^3_w) and the sorbed concentrations as chemical mass per mass of solids (Mc/Ms). Frankly, this difference in metrics can be a source of confusion for prospective modelers, and we adopt the simpler equal unit approach in this example to minimize the obfuscation.

With these changes, the new set of equations for the modified PRAM example follows.

Ship compartment:

$$V_1 \frac{dc_1}{dt} = W + \frac{E_{1,2}A_{1,2}}{L_{1,2}}(c_2 - c_1) - kV_1 c_1 \qquad (2.1)$$

---

[*] We use a double subscript (ds) for desorption to avoid confusion with a later K_d parameter that is the commonly used equilibrium partitioning coefficient.

[†] In this book, except in numerical examples, we use generic unit notations: M (mass), L (length), L^2 (area), L^3 (volume). For example, chemical mass per total volume is represented as Mc/L^3_T, chemical mass per volume of water is Mc/L^3_w, and mass of solids per mass of chemical is Ms/Mc.

Water column compartment (c_{d3} replaces original $F_d c_3$):

$$V_2 \frac{dc_2}{dt} = \frac{E_{1,2}A_{1,2}}{L_{1,2}}(c_1 - c_2) + \frac{E_{2,3}A_{2,3}}{L_{2,3}}(c_{d3} - c_2) + \frac{E_{2,4}A_{2,4}}{L_{2,4}}(c_4 - c_2) - Q_2 c_2 - kV_2 c_2$$

$$(2.2)$$

Sediment compartment (two new equations):

$$V_3 \frac{dc_{d3}}{dt} = \frac{E_{2,3}A_{2,3}}{L_{2,3}}(c_2 - c_{d3}) - kV_3 c_{d3} + k_{ds}V_3 c_{p3} - k_s V_3 c_{d3} \qquad (2.3)$$

$$V_3 \frac{dc_{p3}}{dt} = k_s V_3 c_{d3} - k_{ds}V_3 c_{p3} - kV_3 c_{p3} \qquad (2.4)$$

Air compartment:

$$V_4 \frac{dc_4}{dt} = \frac{E_{2,4}A_{2,4}}{L_{2,4}}(c_2 - c_4) - Q_4 c_4 \qquad (2.5)$$

where c_1, c_2, and c_4 will continue to represent dissolved chemical because of our simplifying assumption of no solids in those compartments. Thus, our modified problem includes four compartments (as before) with one state variable[*] (dissolved chemical) applicable in all of those compartments, and a second state variable (particulate chemical) applicable only in the sediment compartment. Two new rate constants (k_{ds} and k_s) have also been introduced.

How do we assign values to our k_{ds} and k_s rate constants to make the problem equivalent to the $F_d = 0.1$ that we originally used? Let's assume for a moment that our sorption/desorption process is operating within an enclosed, completely mixed, volume with no inflow or outflow (a batch reactor) and no other kinetic processes are present. Under this scenario, we can write the two mass balance equations as

$$\frac{dc_d}{dt} = k_{ds}c_p - k_s c_d \qquad (2.6)$$

$$\frac{dc_p}{dt} = k_s c_d - k_{ds}c_p \qquad (2.7)$$

---

[*] A state variable describes the state of a dynamic system, for example, the concentration of a chemical at a given time. We use the state variable term occasionally in this book for both dynamic and steady-state systems. A state variable denotes a *form* of a chemical, e.g., total versus dissolved or phosphate phosphorus versus algal phosphorus, or a particular oxidation state. Therefore, if we had, for example, two state variables being modeled in each of five compartments, we would have $2 \times 5 = 10$ chemical concentrations at any time step.

At steady state, we can solve for the ratio of the dissolved and particulate chemical from either of the above equations as

$$\frac{c_d}{c_p} = \frac{k_{ds}}{k_s}$$

or

$$c_d = \left(\frac{k_{ds}}{k_s}\right) c_p \tag{2.8}$$

By definition, (again, we are here assuming equal units for c_d and c_p)

$$F_d = \frac{c_d}{c_d + c_p}$$

and, substituting from (2.8),

$$F_d = \frac{\left(\frac{k_{ds}}{k_s}\right) c_p}{\left(\frac{k_{ds}}{k_s}\right) c_p + c_p} = \frac{1}{1 + \frac{k_{ds}}{k_s}} \tag{2.9}$$

We also know that the relationship between F_d and k_{ds} and k_s given by Equation (2.9) and developed under the steady-state assumption represents the equilibrium relationship as well. Why? Because in our batch reactor there are no transport bottlenecks because there is no transport. With no transport, there is an infinitely long reaction time between the two chemical phases to reach equilibrium. Therefore, from Equation (2.9) we can assign $F_d = 0.1$ and calculate

$$\frac{k_s}{k_{ds}} = 9$$

which should parameterize the k_{ds} and k_s rate constants so that we can presumably reproduce the PRAM instantaneous equilibrium (F_d-based) example originally presented in Chapter 1, but with a modified set of equations.

We assigned $k_s = 0.09$/day and $k_{ds} = 0.01$/day to satisfy this ratio and ran the GEM under steady-state conditions for the modified PRAM-like model. (All other parameters and loads are the same.) The results for the ship, water column, and air compartments were identical to the Chapter 1 results. However, the dissolved and particulate concentrations in the sediment were, respectively, 1.09×10^{-2} g/m^3 and 6.98×10^{-3} g/m^3 for a total sediment concentration of 1.78×10^{-2} g/m^3, nearly an order of magnitude less than our previous, Chapter 1 total concentration of 1.08×10^{-1} g/m^3. In addition, the *realized* F_d from our new model is

$$F_d = \frac{1.09 \times 10^{-2}}{1.09 \times 10^{-2} + 6.98 \times 10^{-3}} = 0.18$$

—not particularly close to the 0.1 value we thought we were simulating.

What's wrong? The problem with our new model construct gets back to the issue of steady state versus equilibrium. We are at steady state because we ran the model that way. However, with the relatively slow rate constants we assigned ($k_s = 0.09$/day and $k_{ds} = 0.01$/day) even though they satisfy our 9 ratio, the rate constants themselves represent a bottleneck so that equilibrium is not achieved.

The relatively slow sorption rate of 0.01/day is a bottleneck in that the equilibrium sorbed concentration is not achieved. Dissolved chemical arrives in the sediment from the overlying water column. If it then cannot quickly go into the particulate phase before it is diffused back to the water column (and advected away and/or biodegraded), then equilibrium will not be reached. This is an example of steady state, but no equilibrium. (There is an equilibrium, however. Note that the dissolved concentrations in the water column and the sediments are identical. Thus, the diffusion coefficient across the sediment/water column interface is fast enough that the water column and sediment pore water attain equilibrium. This is a case of steady state and equilibrium.)

With this insight, we can remove the kinetic bottleneck in the sediments by simply assigning higher rate coefficients, still with the 9 ratio. To make sure our kinetics were fast enough, we increased each rate constant by four orders of magnitude to $k_s = 90$/day and $k_{ds} = 10$/day. We ran the GEM and, again, the ship, water column, and air compartments had identical results to the Chapter 1 example. However, this time our dissolved and particulate sediment concentrations were, respectively, 1.08×10^{-2} and 9.70×10^{-2}, for a total concentration of 1.08×10^{-1}, identical to our Chapter 1 result. Furthermore, the realized F_d is

$$F_d = \frac{1.08x10^{-2}}{1.08x10^{-2} + 9.70 \times 10^{-2}} = 0.10,$$

as we desired. By removing the sediment kinetic bottleneck, we now have steady-state and equilibrium conditions in the sediments, and have reproduced the Chapter 1 $F_d = 0.1$ scenario with our modified model.

Let's try another experiment. Above, we noted that even with a kinetic bottleneck in the sediment compartment, the sediment pore water and water column concentrations were identical, and thus in equilibrium. The diffusion parameter was apparently fast enough to allow this equilibrium. We reduced the E_{23} parameter from 1.0 m^2/day to 0.001 m^2/day and re-ran the problem. Again, the ship, water column, and air concentrations are identical to previous results. However, this time our dissolved and particulate sediment concentrations are, respectively, 4.30×10^{-3} and 3.91×10^{-2}. Although the realized F_d is 0.1, illustrating continued equilibrium in the sediments (thanks to our faster rate constants), the dissolved sediment concentration is now considerably less than the previous runs (1.08×10^{-2}), illustrating the loss of equilibrium between water column and pore water concentrations. The diffusion across the sediment–water column interface is now a bottleneck.

Finally, just because this is an interesting and hopefully insightful example, we removed all transport and kinetic bottlenecks and made the air compartment and water compartments advective flows identical. All dispersion/diffusion coefficients were set to 1,000 m²/day and the two flow rates at 10,000 m³/day. The sorption and desorption rate constants remain at their fast values. Running the example with these bottlenecks removed and equal flows, we got

$$c_1 \text{ (ship)} = c_2 \text{ (water column)} = c_4 \text{ (air)} = c_{d3} \text{ (sediment)} = 0.031 \text{ g/m}^3$$

$$c_{p3} \text{ (sediment)} = 0.280 \text{ g/m}^3$$

We see equilibrium among dissolved chemical in all four compartments.[*] In the sediment, the F_d continues to be the equilibrium value of 0.1.

In the remainder of this chapter, we will refer to this version of the PRAM-like model as the k_s/k_{ds} version.

2.1.2 K_d ALTERNATIVE

Our third alternative formulation of the PRAM-like model is to use an equilibrium partition coefficient-approach. Like the Chapter 1 F_d version, this alternative will result in four equations. (The k_s/k_{ds} version involved five equations.) Unlike the Chapter 1 F_d version, we will model only dissolved (not total) PCB in the sediment compartment.

The K_d partition coefficient is a widely used parameter that quantifies chemical partitioning into dissolved and sorbed phases under equilibrium conditions. An experiment to estimate K_d for a specific chemical proceeds by adding solids to a completely-mixed container containing dissolved chemical. Over time, samples of both the solids and water are collected and the dissolved and sorbed concentrations are measured. These data are then plotted as c_p versus c_d (isotherm plot). After equilibrium is reached, the slope of the data indicate the equilibrium partitioning relationship. This slope is the partition coefficient K_d:

$$K_d = \frac{c_p}{c_d} \tag{2.10}$$

With the dissolved and sorbed concentrations having units of, respectively, Mc/L³w and Mc/Ms; K_d has units of L³w/Ms. (Unlike the k_s/k_{ds} version, where for simplicity we used equal units of Mc/L³T for both dissolved and sorbed concentrations, we are here reverting to the more common units. Box 2.1 describes the kinetic k_s/k_{ds} formulation using the more common units.) K_d values are chemical-specific and widely available in the literature.

With these units for the dissolved and sorbed phases, the total concentration is

$$c_t = c_d\theta + c_p BD \tag{2.11}$$

---

[*] Atmospheric chemistry is very complex and one would not expect a dissolved concentration in air to be the same as in a water body under any reasonable circumstances. Nonetheless, we are using a highly simplified description of air in this example, and, under the assumptions of our example, they should be the same.

where

$c_t =$ total concentration (Mc/L³total)
$\theta =$ water content (L³w/L³total)
$BD =$ solids bulk density (Ms/L³total)

and K_d is then related to the equilibrium F_d as

$$F_d = \frac{c_d\theta}{c_d\theta + c_p BD} \tag{2.12}$$

and substituting $c_p = K_d c_d$ from (2.10) into (2.12) results in

$$F_d = \frac{c_d\theta}{c_d\theta + K_d c_d BD} = \frac{\theta}{\theta + K_d BD} \tag{2.13}$$

The particulate fraction F_p is then

$$F_p = 1 - F_d = 1 - \frac{\theta}{\theta + K_d Bd} = \frac{K_d BD}{\theta + K_d BD} \tag{2.14}$$

BOX 2.1   KINETIC SORPTION/DESORPTION MODEL USING CONVENTIONAL UNITS

If we wanted to model the sorption and desorption processes explicitly, as in the previous k_s/k_{ds} version but now, using conventional units, the two mass balance equations for our batch reactor, analogous to Equations (2.6) and (2.7), become

$$\frac{d(c_d\theta)}{dt} = k_{ds}c_p BD - k_s c_d\theta \tag{2.15}$$

$$\frac{d(c_p BD)}{dt} = k_s c_d\theta - k_{ds}c_p BD \tag{2.16}$$

and, at steady state (and equilibrium) from either equation we can solve for

$$\frac{c_p}{c_d} = \frac{k_s\theta}{k_{ds}BD} = Kd \tag{2.17}$$

To relate K_d to the F_d parameter, F_d is, from (2.13) and (2.17),

$$F_d = \frac{\theta}{\theta + K_d BD} = \frac{\theta}{\theta + \left(\dfrac{k_s\theta}{k_{ds}BD}\right)BD} = \frac{1}{1 + \dfrac{k_s}{k_{ds}}} \tag{2.18}$$

just as we found in the simple example when identical units for c_d and c_p were used.

We can now formulate our PRAM-like alternative model using the partition coefficient-approach in the sediment compartment. We now have a single state variable (*dissolved* PCB) in all four compartments. The mass balance equation for PCB in the ship compartment is again identical to both previous alternatives.

$$V_1 \frac{dc_1}{dt} = W + \frac{E_{1,2}A_{1,2}}{L_{1,2}} (c_2 - c_1) - kV_1c_1 \tag{2.19}$$

The water column compartment now assumes c_3 is dissolved chemical

$$V_2 \frac{dc_2}{dt} = \frac{E_{1,2}A_{1,2}}{L_{1,2}} (c_1 - c_2) + \frac{E_{2,3}A_{2,3}}{L_{2,3}} (c_3 - c_2) + \frac{E_{2,4}A_{2,4}}{L_{2,4}} (c_4 - c_2) - Q_2c_2 - kV_2c_2 \tag{2.20}$$

The sediment compartment now involves a retardation sink R that is a function of the partition coefficient and is described in Chapter 4. For purposes of this example, we simply assert that the retardation sink appears in the mass balance equation as shown below

$$R_3V_3 \frac{dc_3}{dt} = \frac{E_{2,3}A_{2,3}}{L_{2,3}} (c_2 - c_3) - kV_3c_3 \tag{2.21}$$

where

$$R_3 = 1 + \frac{BD_3K_d}{\theta_3} \tag{2.22}$$

and
$\quad BD_3$ = sediment bulk density (g of solids/m³ total)
$\quad \theta_3$ = sediment porosity (m³ water/m³ total)

Finally, the air compartment mass balance equation is as in both previous alternatives

$$V_4 \frac{dc_4}{dt} = \frac{E_{2,4}A_{2,4}}{L_{2,4}} (c_2 - c_4) - Q_4c_4 \tag{2.23}$$

To calculate R_3, we assumed that the sediment bulk density is 2.5×10^6 g/m³ and the porosity is 0.5. Then, from (2.18) we can calculate a K_d to yield $F_d = 0.1$ as $K_d = 1.8 \times 10^{-6}$ m³/g. (In typical practice, one has a chemical-specific K_d value that *determines* F_d, not the other way around. Here, we are reversing this process to maintain our earlier assumption of $F_d = 0.1$.)

With this value of R_3 and all other parameters and loadings as used previously (in the original Chapter 1 problem), we ran the GEM at steady state for this alternative formulation. The steady-state concentrations are

c_1 (ship) $= 6.23 \times 10^{-2}$ g/m^3

c_2 (water column) $= 1.09 \times 10^{-2}$ g/m^3

c_3 (sediment) $= 1.09 \times 10^{-2}$ g/m^3

c_4 (air) $= 6.99 \times 10^{-3}$ g/m^3

The concentrations in the ship, water column, and air compartments are again identical to those from both previous alternative formulations of this problem. The sediment compartment concentration (dissolved chemical) is identical to the dissolved concentration we found in our kinetic sorption/desorption approach.

The observant reader might note from Equation (2.21) that the R_3 term is simply moot at steady state, because the left side of the equation is 0, and wonder if we are pulling some trick. Indeed, the retardation phenomenon ceases under steady-state conditions. Nonetheless, the retardation approach to this problem is legitimate and certainly needed for dynamic problems. Running this problem under a dynamic scenario, the resulting time series for dissolved chemical in compartment 3 is identical to that time series using the kinetic sorption/desorption alternative or the total chemical (dissolved + sorbed) time series from the Chapter 1 F_d approach after multiplying that time series by F_d (0.1).

In the remainder of this chapter, we will refer to this version of the PRAM-like model as the K_d version.

2.2   GEM INPUT FILES

We now begin the GEM user's guide. The guide is essentially a description of the input files, much of which is self-explanatory. Once the input files are populated, the GEM is executed and writes output files containing simulated concentrations within the compartments as well as several diagnostic files.

All input and output files are comma-separated text files—sometimes called flat files. We have purposely avoided use of a formal database system to hold input and output data to maintain user flexibility. The input files may be created by any text editor or, most conveniently, by using a spreadsheet and saving the results in text, comma-separated-value (.csv) format.* The GEM accesses these files using a sequential access method, i.e., reading data down the records and across each record's fields in a sequential manner.

Table 2.1 maps the input files required for the various uses of the GEM. For example, a dynamic, nonlinear problem in *Environmental System* mode may involve inputs from 18 different files. That many files may seem like overkill (certainly we could put all the data in a single file), but the information has been allocated among the various files so that each file involves only the information needed for those inputs. In addition, most of the inputs (loadings and/or parameters) may be changed dynamically during run time using the shell procedure, discussed later. It is expected

---

* A word of caution is needed. Some of the input files have records with varying numbers of fields. These "ragged-right-edge" files, when saved as .csv files by some spreadsheets (e.g., Microsoft Excel™), will append a series of commas (,,,,,,,) to the right of the shorter records. These commas that do not separate actual data will be read as containing 0 between the commas, generating runtime errors. They must be deleted prior to running the GEM.

TABLE 2.1
Required Input Files

Input File (*.csv)	Environmental System Mode				Equation Solver Mode			
	Steady-State		Dynamic		Steady-State		Dynamic	
	Linear	Nonlinear	Linear	Nonlinear	Linear	Nonlinear	Linear	Nonlinear
Control					✓			
SVCompMap								
Compartments[a]								
Interfaces[a]								
Flows[a]								
ECoefficients[a]								
Flowand								
Emultipliers[a]			✓					
Loads[a]								
VolumeSrcSnks								
AreaSrcSnks								
LinearKdand								
TempCoef[a]								
Boundary[a]								

Initial					✓	
Fofc[b]	✓		✓		✓	✓
Dfdc[b]	✓		✓		✓	✓
Rofc[b]	✓				✓	
dRdc[b]	✓				✓	
Guess					✓	✓
AEqnSolver	✓	✓	✓	✓	✓	
BEqnSolver	✓	✓	✓	✓		
RVEqnSolver	✓	✓				
InitialEqnSolver	✓	✓				
GuessEqnSolver	✓		✓			

[a] File can be static or created by shell procedure.

[b] File is created only by shell procedure.

that only a small subset of these inputs would need to vary in any given application; therefore, allocating them among different files facilitates their dynamic updating.

The files shown in Table 2.1 as required for a particular type of GEM run are just that: *required*. For example, for a dynamic, linear run under *Environmental System* mode, 13 files are required regardless of the parameterization of a specific problem. These files must exist, but they may be essentially empty, except for the descriptive header records and, for some, the -999 last record to signify end of file. Otherwise, a run-time error will result. (For example, you may not have dispersion coefficients in your problem, but the ECoefficients.csv file must exist.)

In general, the files are created by the user a single time prior to a run, and are not changed during the run. We call these static files. Alternatively, most of the input files can also be altered dynamically during runtime by the shell procedure. These are called dynamic files. The name does not necessarily mean changes in time, although time changes are examples of dynamic files. In addition to updating files in time, the shell procedure may be used to provide external functionality to the GEM, for example, user-specified nonlinear source and sink terms for a steady-state nonlinear problem. These nonlinear source and sink terms would be dynamically evaluated during runtime by the shell procedure during iterations of the nonlinear solution algorithm (Newton's method). Thus, the shell procedure and the creation of dynamic files can occur for either steady-state or dynamic problems. The shell procedure is discussed in detail later. First, we discuss the static files, then the dynamic files.

2.2.1 STATIC INPUT FILES

We illustrate the GEM input files using our three versions of the PRAM-like example, i.e., the F_d version, k_s/k_{ds} version, and K_d version.

Control.csv—Common to both *Environmental System* and *Equation Solver* modes is the Control.csv file containing GEM run instructions and overall system parameters. Table 2.2 is a Control.csv file illustrating inputs for the various PRAM-like models. The three primary sources of information are the first three entries, *Environmental System* or *Equation Solver* mode, linear or nonlinear, and steady-state or dynamic. Depending on these entries, many of the remaining entries may be irrelevant, but all these records must be present in the file. For example, this particular file specifies *Environmental System* mode for a linear steady-state problem. Any data pertaining to dynamic, nonlinear, or *Equation Solver* problems are read, but are irrelevant to the run.

Each record for input data is preceded by a single record describing the input field(s) in the next record. Your file must follow the same format, i.e., a populated, single* record occurs between each data field record. If the data record is not relevant (e.g., the time step for a steady-state problem), what is in the data record and

---

* Some of the description records wrap to the next line in the table. It should be understood that these are single records, however. (There are 72 records.)

TABLE 2.2
Control.csv for PRAM-Like Model

```
Environmental System or Equation Solver mode (EnvS, EqnS)?
EnvS

Linear problem? (Y,N)
Y

Steady-state simulation (else dynamic)? (Y,N)
Y

If EqnS, enter number of equations.
4

If EnvS, maximum number of state variables, number of
nonboundary compartments, number of dummy + boundary
compartments.
2,4,2

If EnvS, maximum number of adjacent compartments.
5

If EnvS, enter MixLengthOption.
2

If EnvS, enter around-compartments flow balance tolerance (%).
1

If dynamic, enter 1 for FT,Euler's method, 2 for CT,MacCormack's
method, 3 for BT,Implicit method, 4 for CT,Crank-Nicolson.
3

If dynamic, enter time step in days.
1

If dynamic, enter number of time steps.
200

If dynamic, enter time step at which to output intermediate
diagnostic files.
1

If dynamic and EnvS, enter option (1,2, or 3) for results
output.
3

If write option above = 1, enter time step at which to print
profile.
200

if write option above = 2, enter compartment number and time
step print interval.
3,2
```

Continued

TABLE 2.2 (continued)

Control.csv for PRAM-Like Model

If nonlinear and dynamic, is your R matrix nonlinear, i.e., a function of concentrations? (Y,N)
Y

If nonlinear, are analytical derivatives for Jacobian available (Y,N)?
Y

If no analytical derivatives, enter finite difference multiplier for derivative estimation.
0.1

If nonlinear, would you like to save Newton algorithm diagnostics to NewtonInfo.dng file? (Y,N)
Y

If nonlinear, will zero or negative concentrations present numerical problems during Newton method iterations? (Y,N)
Y

If nonlinear, enter reasonable minimum and maximum values for concentrations.
0.001,2

If nonlinear, enter Newton's method stopping criterion.
1E-10

If nonlinear, enter (1) MaxNewtonIteration, (2) MaxSamplingIteration, (3) NumSamples.
20,10,10

If nonlinear, are you providing nonlinear f(c) source/sink inputs with user-provided fofc.exe and shell functionality? (Y,N)
Y

if nonlinear and dynamic, are you providing nonlinear R(c) inputs with user-provided Rofc.exe and shell functionality? (Y,N)
Y

If dynamic and EnvS, do you want to use shell functionality to update Compartment.csv? (Y,N)
N

If dynamic and EnvS, do you want to use shell functionality to update Interfaces.csv? (Y,N)
N

If dynamic and EnvS, do you want to use shell functionality to update Flows.csv? (Y,N)
N

TABLE 2.2 (continued)
Control.csv for PRAM-Like Model

```
If dynamic and EnvS, do you want to use shell functionality to
update ECoefficients.csv? (Y,N)
N

If dynamic and EnvS, do you want to use shell functionality to
update FlowandEMultipliers.csv? (Y,N)
N

If dynamic and EnvS, do you want to use shell functionality to
update LinearKdandTempCoef.csv? (Y,N)
N

If dynamic and EnvS, do you want to use shell functionality to
update Boundary.csv? (Y,N)
N

If dynamic and EnvS, do you want to use shell functionality to
update Loads.csv? (Y,N)
N

If dynamic and EqnS, do you want to use shell functionality to
update AEqnSolver.csv? (Y,N)
N

If dynamic and EqnS, do you want to use shell functionality to
update BEqnSolver.csv? (Y,N)
N

If dynamic and EqnS, do you want to use shell functionality to
update RVEqnSolver.csv? (Y,N)
N
```

preceding record do not matter but the records must be populated. (In subsequent input files, we use -999 to denote irrelevant data inputs. That could also be used here.) The records shown in Table 2.2 are the same for every GEM problem; only the specific input values vary among applications. Beyond these caveats, the inputs in Control.csv are either self-explanatory or will be covered in subsequent sections.

2.2.1.1 Environmental System Mode Input Files

We first describe the *Environmental System* mode static input files, followed by the *Equation Solver* static input files. These are described in the order in which they are listed in Table 2.1.

SVCompMap.csv—The SVCompMap.csv information file maps the state variables to the compartments. Table 2.3 is the SVCompMap.csv file for the k_s/k_{ds} version of the PRAM-like model. This file has MaxNSV + 1 records, where MaxNSV is the maximum number of state variables (chemical forms) entered in the Control.csv file. One does not have to model the same number of state variables in all compartments. (In the k_s/k_{ds} example, we model one state variable [dissolved PCB] in compartments

---

TABLE 2.3

SVCompMap.csv

```
SV,Comp 1,Comp 2,... ,,,, Comp NT
1,1,1,1,1,0,0
2,0,0,1,0,0,0
```

1, 2, and 4, and two state variables [dissolved and particulate PCBs] in the sediment compartment 3.) MaxNSV is the maximum number of state variables modeled among all compartments. The first record is a description header record.

The first field in the input records is the state variable number. State variables are integer- and sequentially numbered from 1 to MaxNSV. The arrangement of your various state variables relative to each other does not matter, but when you populate the SVCompMap.csv file, the numbering of your state variables assigns an internal GEM mapping. For example, in Table 2.3, we treated dissolved PCB as state variable 1 and particulate PCB as state variable 2. (We could have switched this around.) Within the GEM, and in output files, dissolved PCB is then SV 1 and particulate PCB is then SV 2.

The next NT fields represent the NT compartments (total number of compartments) and contain 1s if that state variable is relevant to that compartment and 0s if not relevant. These NT fields are treated as corresponding to the sequentially numbered compartments from 1 to NT. For example, a 1 in field 2 means that the state variable in field 1 is relevant to compartment 1, a 2 in field 3 means relevance to compartment 2, and so on.

You must not enter a 1 in any boundary or dummy compartment (e.g., compartments 5 and 6 for the PRAM-like example) even though the fields include those boundary or dummy compartments. (This was done to facilitate coding.) An error will be generated if a 1 is entered into any dummy or compartment field.

Compartments.csv—Compartments.csv provides the compartment-specific information. Table 2.4 is the Compartments.csv file for the K_d version of the PRAM-like model. This file has NT + 1 records where NT is the total number of

---

TABLE 2.4

Compartments.csv

```
Compartment,No Adj Compartments,Boundary,X (m),Y (m),Z (m),
Volume (m^3),Temp (deg C),Water Content (L^3w/L^3total),Bulk
Density (gm/cu m)
1,1,0,0,0,0,120000,20,1,-999
2,5,0,0,0,0,1080000,20,1,-999
3,1,0,0,0,0,1800,20,0.5,2.50E+06
4,3,0,0,0,0,1080000,20,1,-999
5,2,2,0,0,0,-999,-999,-999,-999
6,2,1,0,0,0,-999,-999,-999,-999
```

---

compartments, including boundary and dummy compartments. The first record is a description header record.

The compartment field must be sequentially integer-numbered from 1 through NT. The individual compartments may be arranged in any order, 1-, 2-, or 3-dimensional, and there are no restrictions on the compartment numbering scheme vis-á-vis the arrangement of the compartments relative to each other. Thus, for example, compartment 1 does not have to be adjacent to or even anywhere near compartment 2. In addition, compartment numbering is arbitrary with regard to the type of compartment (dummy, boundary, modeled; see below).

Regarding individual compartment geometry, a compartment can have any shape so long as the cumulative volume of the compartments comprises the spatial domain that one is modeling. If interfacial areas between compartments or mixing length distances are relevant to your model (e.g., dispersive or diffusive mass transport), your choice of compartment volume and shape must of course also consider the practicality of quantifying these parameters. For practical purposes, most users will probably use the right rectangular prism* geometry for compartments. (In the PRAM-like example, compartments 1, 3, and 4 are right rectangular prisms; compartment 2 is not.)

The boundary field is an integer indicator that specifies whether a compartment is a boundary or dummy compartment or not and, if so, what type of boundary condition applies. A dummy compartment is not a boundary or modeled compartment (i.e., a compartment for which concentrations are simulated), but gives a place for flows to go to or from. The following indicators are used:

Value = 0, compartment is modeled, i.e., not a boundary or dummy
Value = 1, dummy compartment
Value = 2, fixed concentration boundary compartment
Value = 3, zero gradient boundary compartment
Value = 4, linear gradient boundary compartment

The three location fields, X (m), Y (m), and Z (m), are the coordinates of the compartment centroid. If these fields are populated (see following section on Interfaces. csv), you can assign your datum coordinates (0,0,0) from which distances to all other compartments are measured to any compartment centroid. (Negative coordinates relative to the datum compartment centroid are fine.) Obviously, if you have only 1-dimensional data, you need only populate one of those fields. For 2-dimensional data, two fields must be populated. Again, -999 is used to indicate irrelevant data.†

Interfaces.csv—Interfaces.csv gives the compartment interfacial information. Table 2.5 is the Interfaces.csv file for all versions of the PRAM-like model. This file has NT + 1 records where NT is the total number of compartments, including boundary and dummy compartments. The first record is a description header.

---

* A solid figure bounded by six faces where each face is a rectangle and each pair of adjacent faces meets in a right angle.

† There are numerous checks in the GEM on whether an input value of -999 is used in a calculation. If it is, the run terminates and an error message is generated.

TABLE 2.5

Interfaces.csv

```
Compartment,Adj Compartment,Interface Area (m^2),Centroid to
Interface Distance (m),Mixing Length (m),alpha,Adj
Compartment,Interface Area (m^2),Centroid to Interface
Distance (m),Mixing Length (m),alpha,Adj Compartment,
Interface Area (m^2),Centroid to Interface Distance (m),
Mixing Length (m),alpha
1,2,1800,-999,1,1.0
2,1,1800,-999,1,1.0,3,12000,-999,0.1,1.0,4,18000,-
999,10,1.0,5,-999,-999,-999,1.0,6,-999,-999,-999,1.0
3,2,12000,-999,0.1,1.0
4,2,18000,-999,10,1.0,5,-999,-999,-999,1.0,6,-999,-999,
-999,1.0
5,2,-999,-999,-999,1.0,4,-999,-999,-999,1.0
6,2,-999,-999,-999,1.0,4,-999,-999,-999,1.0
```

The first (compartment) field must be sequentially integer-numbered from 1 to NT. The next five fields are repeated for each adjacent compartment and are self-explanatory, except for "alpha." Alpha is the spatial weighting parameter described in Chapter 3. Adjacent compartment numbers can be in any order; however, *the ordering must be consistent among the input files:* Interfaces.csv, Flows.csv, and ECoefficients.csv. It is suggested that the user build one of these files and then use that file as a template for the remaining two.

An inefficiency in this file requires you to enter the same information twice. The data entered for the record corresponding to compartment i and adjacent compartment j must also be entered in the record for compartment j and adjacent compartment i. This is admittedly a slight nuisance, but this design was adopted to facilitate coding and run speed. Two other files to be discussed (Flows.csv and ECoefficients.csv) also have this inefficiency.

Regarding the several fields in Compartments.csv. and Interfaces.csv that contain spatial location or compartment length data, namely X (m), Y (m), and Z (m) in Compartments.csv and Centroid to Interface Distance (m) and Mixing Length (m) in Interfaces.csv, these data are used as follows:

To set the mixing distance over which dispersion/diffusion across compartment interfaces occurs—As described in Section 5.3.2, in calculating dispersion/diffusion mixing length, the appropriate mixing length is generally considered to be the distance over which a concentration gradient exists between two compartments. The following MixLengthOption choices apply:

MixLengthOption = 1 uses the X (m), Y (m), Z (m) fields in Compartments. csv. The distance between the centroids of compartments i and j is calculated and used as the mixing length for dispersion/diffusion between i and j. For example, suppose you are simulating a system where you would expect a continuous concentration gradient among the compartments and the compartments are convex spaces. A convex space is a space where all points along a line connecting any two points within the space also are within the space. For example, the line connecting two points on opposite sides of the "hole" in a doughnut passes through the doughnut hole, which is outside the doughnut space. Thus a doughnut is not a convex space. (Although the doughnut hole itself is a convex space.) Another example is the water column compartment in our PRAM-like example that wraps around the ship compartment on three sides. The centroid of the water column compartment is not within the water column compartment, and one would not want to use MixLengthOption = 1 here.

MixLengthOption = 2 uses the Mixing Length (m) fields in Interfaces.csv. Specify this option when (1) your compartments are not convex, and/or (2) the distances between compartment centroids do not accurately reflect the mixing lengths. For example, if your compartments are physically well mixed and concentrations vary only in very close proximity across the interfaces, a mixing length between compartment centroids may significantly overstate the true mixing length and MixLengthOption = 2 is appropriate.

To set up linear gradient boundary conditions—As described in Section 3.6.3, a linear gradient boundary condition between an interior compartment i and one or more adjacent, boundary compartments uses the distances between centroids of those compartments and is implemented by the GEM using the X (m), Y (m), Z (m) fields in Compartments.csv. This is the only current option for the linear gradient boundary condition and the user must ensure that these distances are appropriate given their compartmental arrangement, i.e., non-convex compartments may result in inappropriate distances.

To evaluate wiggle criterion—As described in Section 5.3.2, the wiggle criterion uses distances between compartment centroids to ensure that compartments are not so large as to induce the numerical errors of solution oscillation and/or non-positive results. This criterion is implemented by the GEM using the X (m), Y (m), Z (m) fields in Compartments.csv. Again, this is the only current option and the user must ensure that these distances are appropriate given their compartmental arrangement, i.e., non-convex compartments may result in inappropriate distances.

To determine the spatial differencing parameters, α_{ij}, when compartments are different sizes and a central difference is desired—As described in Section 5.3.2, when two adjacent compartments are different sizes and a central differencing method is used, the α_{ij} parameter needs to be biased (weighted) toward the compartment whose centroid is the closest to the compartment interface, i.e., not equal to 0.5. These weighted α_{ij} parameters are calculated using the Centroid to Interface Distance (m) data in Interfaces.csv.

As a convenience in output files so that a user can plot concentrations against distance if desired—The output file ProfileSnapshot.csv is a snapshot of the concentrations in all compartments at a given time step or for a steady-state simulation. This output file repeats the X (m), Y (m), Z (m) fields from Compartments.csv as a user convenience for subsequent graphical displays.

TABLE 2.6

Flows.csv

```
State Variable,Compartment,Adj Compartment,Flow (cmd),Adj
Compartment,Flow (cmd),Adj Compartment,Flow (cmd)
1,1,2,0
1,2,1,0,3,0,4,0,5,10000,6,-10000
1,3,2,0
1,4,2,0,5,100000,6,-100000
1,5,2,-10000,4,-100000
1,6,2,10000,4,100000
2,1,2,0
2,2,1,0,3,0,4,0,5,10000,6,-10000
2,3,2,0
2,4,2,0,5,100000,6,-100000
2,5,2,-10000,4,-100000
2,6,2,10000,4,100000
```

If some of these uses are irrelevant to your application, you can populate those data fields with -999.

Flows.csv—Flows.csv gives the inter-compartmental advective flow rates that may vary by state variable. (For example, multi-phase flow regimes such as contaminant fate and transport in an unsaturated zone of soil may have water-related and air-related flow regimes.) Table 2.6 is the Flows.csv file for the k_s/k_{ds} version of the PRAM-like model. This file has MaxNSV*NT + 1 records. The first record is a description header.

There are MaxNSV blocks of NT records. These MaxNSV blocks must be in sequential order of state variable numbering in the first field from 1 to MaxNSV; that is, the first NT records (after the header) relate to state variable 1, the second set of NT records to state variable 2, and so on. Within each block, the records are also sequentially numbered by compartment from 1 to NT in field 2.

The next two fields are repeated for each adjacent compartment. Adjacent compartment numbers can be in any order; however, *the ordering must be consistent among the input files: Interfaces.csv, Flows.csv and ECoefficients.csv. Flows (m³/day) are entered as positive values if the flow is entering the compartment in field "Compartment" and negative if it is leaving.* For example, in the fifth record (pertaining to compartment 4), a flow of 100,000 m³/day is entering compartment 4 from boundary compartment 5 and thus is positive. The flow going to dummy compartment 6 is leaving compartment 4 and so is negative.

ECoefficients.csv—The ECoefficients.csv file provides the inter-compartmental interface dispersion/diffusion rate constants. These data are state variable-dependent, for the reason that diffusion, unlike macro dispersion,* is a chemical-specific

---

* Macro dispersion refers to large-scale mixing mechanisms such as tidal flow reversals or other mixing phenomena related to non-spatially uniform flow patterns.

TABLE 2.7

ECoefficients.csv

```
State Var,Compartment,Adj Compartment,E (m^2/day),Adj
Compartment,E (m^2/day),Adj Compartment,E (m^2/day)"
1,1,2,1.0
1,2,1.0,1,3,1.0,4,100,5,-999,6,-999
1,3,2,1.0
1,4,2,100,5,-999,6,-999
1,5,2,-999,4,-999
1,6,2,-999,4,-999
2,1,2,-999
2,2,1,-999,3,-999,4,-999,5,-999,6,-999
2,3,2,0
2,4,2,-999,5,-999,6,-999
2,5,2,-999,4,-999
2,6,2,-999,4,-999
```

parameter usually determined as a function of molecular weight. If your coefficients are all macro dispersions, you just need to enter the same data for each state variable block. If they are diffusion-related, the flexibility exists in this file to make them state variable-specific

Table 2.7 is the ECoefficients.csv file for the k_s/k_{ds} version of the PRAM-like model. This file has MaxNSV*NT + 1 records. The first record is a description header. The MaxNSV blocks must be in sequential order of state variable numbering in the first field from 1 to MaxNSV; that is, the first NT records (after the header) relate to state variable 1, the second set of NT records to state variable 2, and so on. Within each block, the records are also sequentially numbered by compartment from 1 to NT in field 2.

In addition to the inefficiency from entering the same data twice previously mentioned, an additional inefficiency exists in this file because you must enter NT records for each state variable, regardless of whether that state variable is relevant to all NT compartments. This design was also adopted to facilitate coding and run speed. If a state variable is not relevant to a compartment, simply enter -999 for the dispersion/ diffusion coefficient. For example, in Table 2.3, state variable 2 (particulate PCB) is relevant only to compartment 3. Thus, all dispersion/diffusion coefficients are -999 in each of the second block NT records except for the record corresponding to compartment 3. In addition, if a dispersion/coefficient is not relevant (needed) for a particular interface (even if the state variable is relevant), also enter -999. For example, in Table 2.7, we are not simulating dispersion/diffusion from compartment 2 to compartments 5 or 6. Those values are also -999.

Beginning with field 3, there are two fields per adjacent compartment. The first of these is the adjacent compartment number. Adjacent compartment numbers can be in any order; however, *the ordering must be consistent among the input files: Interfaces.csv, Flows.csv and ECoefficients.csv*. The second is the

---

TABLE 2.8

FlowandEMultipliers.csv

```
SV, Equation Compartment, Compartment Containing Multiplier,
Interface I, Interface J, Eij multiplier, Qij multiplier
1,2,3,2,3,0.1,1
1,3,3,2,3,0.1,1
-999
```

---

diffusion/dispersion coefficient (m²/day) pertaining to the interface between the field 2 compartment and the adjacent compartment.

FlowandEMultipliers.csv—This file allows the user to specify arbitrary multipliers on the flow and dispersion parameters. For example, recall our Chapter 1 F_d-based version of the PRAM-like model. Our state variable there was total chemical, and the F_d coefficient (fraction dissolved) appeared in the dispersion terms of Equations (1.4) and (1.6) to modify the c_3 state variable (total chemical in compartment 3, the sediments). This F_d- multiplier accounts for the fact that only the dissolved fraction of the chemical diffuses across the sediment–water column interface. The F_d dispersion coefficient multiplier is specified via the FlowandEMultipliers.csv file for this example. User-specified multipliers on the flow terms are entered in an analogous manner to the dispersion multipliers described here. These arbitrary multipliers represent any multipliers *other* than the standard parameters in the generic transport terms for dispersion and/or flow described in Chapter 3.

Table 2.8 is the FlowandEMultipliers.csv file for the F_d version of the PRAM-like example from Chapter 1. There are a user-specified number of records. The first record is a description header. The last record must be -999 to signify the end of the file. If there are no non-standard flow and/or dispersion multipliers, the file still must exist but will consist of only these two records.

Each of the intermediate records contains information that will modify one flow and/or one dispersion parameter in one equation. For example, our Chapter 1 F_d-based example involves the F_d multiplier for a single state variable (c_3) but in two equations. Considering the dynamic form of those equations, the two affected equations are (1.4) and (1.6). Therefore, we must have two records. If we also had a (single) flow multiplier in those equations, we could modify this flow with the same two records. If we had a flow multiplier in another equation, we would need another record. If we had two flow and/or dispersion multipliers in a single equation, we would need two records.

The structure of the intermediate records is described below. It will help the reader to follow this explanation by having the F_d version equations available and they are repeated below.

$$V_1 \frac{dc_1}{dt} = W + \frac{E_{1,2} A_{1,2}}{L_{1,2}} (c_2 - c_1) - kV_1 c_1 \qquad \text{(1.2 repeated)}$$

$$V_2 \frac{dc_2}{dt} = \frac{E_{1,2}A_{1,2}}{L_{1,2}}(c_1 - c_2) + \frac{E_{2,3}A_{2,3}}{L_{2,3}}(F_d c_3 - c_2) + \frac{E_{2,4}A_{2,4}}{L_{2,4}}(c_4 - c_2) - Q_2 c_2 - kV_2 c_2$$

(1.4 repeated)

$$V_3 \frac{dc_3}{dt} = \frac{E_{2,3}A_{2,3}}{L_{2,3}}(c_2 - F_d c_3) - kV_3 c_3 \qquad\qquad (1.6 \text{ repeated})$$

$$V_4 \frac{dc_4}{dt} = \frac{E_{2,4}A_{2,4}}{L_{2,4}}(c_2 - c_4) - Q_4 c_4 \qquad\qquad (1.8 \text{ repeated})$$

The first field is the relevant state variable. It is assumed here that any dispersion or flow multipliers will occur within the A coefficient matrix only within the submatrix that is relevant to this state variable. (We cannot envision the need to modify a transport term involving transport *across* state variables. Transport across compartmental interfaces is relevant to a common state variable. Once that state variable is transported into/out of a compartment, it then may undergo kinetic changes into another state variable, but that transformation does not affect the transport terms.) The structures of the A matrix and its submatrices are explained in Chapter 4.

The second field is the number of the compartment whose equation involves the multiplier. (This gives the row number of the A matrix.) The third field is the number of the compartment involving the term in that equation containing the multiplier. (This gives the column number of the A matrix.) The fourth and fifth fields are the compartment numbers of the two adjacent compartments involving the multiplier. For example, to modify $E_{i,j}$, i.e., the dispersion parameter for compartments i and j, these inputs are the i and j compartment numbers. (It should be noted that a given element of the A matrix may involve several intercompartmental dispersion or flow terms, so input pair (3) above is required to specify which gets the multiplier. For example, Equation (1.4) above has three different dispersion parameters that are coefficients of c_2, although none of these involves a multiplier.) Finally, the sixth and seventh fields are the dispersion and flow multiplier values, respectively.

Returning to the inputs for our Table 2.8 example, the second record of that table pertains to the F_d multiplier in Equation (1.4). The term involving the multiplier is

$$\frac{E_{2,3}A_{2,3}}{L_{2,3}}(F_d c_3) .$$

Field 1 of that record is for state variable 1 (this problem has only 1 state variable). Because (1.4) is the mass balance equation for state variable 1 in compartment 2, the entry for field 2 is 2. In that equation, the term containing the dispersion multiplier is relevant to compartment 3, i.e., c_3. Therefore, the entry for field 3 is 3. In that term, the dispersion coefficient modified is between compartments 2 and 3; therefore, fields 4 and 5 are, respectively, 2 and 3. Finally, fields 6 and 7 are the dispersion and flow multipliers, 0.1 and 1, respectively.

Analogously, record 2 pertains to the F_d multiplier in Equation (1.6). The term involving that multiplier is

$$\frac{E_{2,3}A_{2,3}}{L_{2,3}}(-F_d c_3).$$

Field 1 of that record is again for state variable 1. Because (1.6) is the mass balance equation for state variable 1 in compartment 3, the entry for field 2 is 3. In that equation, the term containing the dispersion multiplier is relevant to compartment 3, i.e., c_3. Therefore, the entry for field 3 is 3. In that term, the dispersion coefficient modified is between compartments 2 and 3; therefore, fields 4 and 5 are, respectively, 2 and 3. Finally, fields 6 and 7 are the dispersion and flow multipliers, 0.1 and 1, respectively.

Loads.csv—The Loads.csv file provides the forcing function or compartmental external loadings. This file is used for both steady-state and dynamic problems. Table 2.9 is the Loads.csv file for all versions of the PRAM-like model under either steady-state or dynamic mode when the load does not change (i.e., when we ran the dynamic model out to steady-state). This file has a user-determined number of records. The file consists of three repeating sets of information, each contained within the previous type. The first type of information is the time window relevant to the contained information. The second is the compartment number for the loadings, and the third type is the state variable within the compartment and time window and the state variable-specific loadings. The default loading for any state variable in any compartment in any time period not specified in the Loads.csv file is assumed to be 0.

In Table 2.9, the first record is descriptive record and must be present to begin each time window block of information. The second record has two fields. The first is the time (days) at which the subsequent loadings begin. The second is the time (days) at which the subsequent loadings cease. The third record is also descriptive and must be present to begin each compartment block of information within the time window. The fourth record gives in field 1 the compartment number for the loading. On that same record, the following repeating fields give first the relevant state variable for that compartment and time window, followed by the state variable-specific loading (g/day). In our example, we have a loading of 1,000 g/day for state variable 1 in compartment 1 in the time window. For each compartment record, the state variable numbers and loadings are repeated as needed, followed by a -999 to signify the end of that set of state variable loading information. The fifth record is -999 signifying the end of the compartment-specific records. The sixth record is also -999

TABLE 2.9

Loads.csv

```
TimeStart, TimeEnd
0,1000000
Compartment,SV, SV Load, SV, SV Load, ...
1,1,1000,-999
-999
-999
```

signifying the end of the time window blocks of information. Thus, the -999 flag terminates each of the three types of information.

Let's look at a more complicated Loads.csv file. Suppose we had different loadings in two time windows (0 to 15 days and 10 to 200 days). The GEM software assigns loadings to a TimeStart, TimeStop window if the current simulation time is greater than or equal to TimeStart and less than TimeStop. (Thus, a new loading starting on day 15 would be included in the 15, 200 time window, not the 0, 15 window. The user could equivalently make this more explicit by entering the time windows as, e.g., 0 to 14.999 and 15.0 to 200 days.) Suppose also we had two state variables. Let state variable 1 have a loading of 10 g/day in compartment 1 within the 0- to 15-day window and a loading of 20 g/day in compartment 2 within the same window. Let state variable 2 have a loading of 22 g/day in both compartments 1 and 2 during that time window. For the second time window (15 to 200 days), assume that only state variable 2 in compartment 2 has a non-zero loading, and that loading is 38 g/day. The relevant Loads.csv file is

```
TimeStart, TimeEnd
0,15
Compartment,SV, SV Load, SV, SV Load, ...
1,1,10,2,22,-999
2,1,20,2,22,-999
-999
TimeStart, TimeEnd
15,200
Compartment,SV, SV Load, SV, SV Load, ...
2,2,38,-999
-999
-999
```

Some problems will not involve external loadings, e.g., the forcing function is entirely provided by boundary or initial conditions. Nonetheless, the Loads.csv file must be present. In these cases, however, the appropriate Loads.csv file would consist of only a single record with a single entry, -999.

Finally, recall in the Chapter 1 F_d version of the PRAM-like model that we made a final dynamic run in which the load varied continuously in time, decaying in accordance with $W(t) = 1,000e^{-0.01t}$. In a case like this, inputting the loads through a static Loads.csv file such as above would be problematic. For example, if you were simulating 200 days at a 1-day time step (as we did in Chapter 1), a static Loads.csv file would look as follows for the first 5 days:

```
TimeStart, TimeEnd
0,1
Compartment,SV, SV Load, SV, SV Load, ...
1,1,995.01,-999
-999
TimeStart, TimeEnd
1,2
Compartment,SV, SV Load, SV, SV Load, ...
1,1,985.11,-999
```

```
-999
TimeStart, TimeEnd
2,3
Compartment,SV, SV Load, SV, SV Load, ...
1,1,975.31,-999
-999
TimeStart, TimeEnd
3,4
Compartment,SV, SV Load, SV, SV Load, ...
1,1,965.61,-999
-999
TimeStart, TimeEnd
4,5
Compartment,SV, SV Load, SV, SV Load, ...
1,1,956.00,-999
-999
-999
```

The entire file would need 1,001 records (five for each day plus the final -999 time window-ending flag record). We chose the midpoint of each day to assign the above loadings, e.g., for day 1, $W(1) = 1,000e^{-0.01(0.5)} = 995.01$. You can see that populating such a file is a lot of work for a long simulation. It is much easier to use the shell functionality (described later) to populate the Loads.csv file in such a situation. For example, for day 4, the user-supplied Loads.exe program would write the complete Loads.csv file (updated daily) for our decaying source example as

```
TimeStart, TimeEnd
3,4
Compartment,SV, SV Load, SV, SV Load, ...
1,1,965.61,-999
-999
-999
```

The next two files relate to specifying linear sources and sinks.

VolumeSrcSnks.csv—Volume-based sources and sinks involve a rate constant times a concentration times a compartment volume. The VolumeSrcSnks.csv file provides the volume-based rate constants for sources and sinks and the information to properly locate these terms in the **A** matrix (see Chapter 4).

All sources/sinks in the PRAM-like examples are volume-based. Table 2.10 is the VolumeSrcSnks.csv file for the sources and sinks in the k_s/k_{ds} version of that example. This file has a user-determined number of records. The first record is a description header. There is one subsequent record for each volume-based source or sink in the model. The -999 record flag is used to signify the end of the file.

Each data record contains the information for only a single source or sink, i.e., a single loss or gain to a single chemical (state variable) in a single compartment. If a loss (gain) to one state variable represent a gain (loss) to another state variable, these two source and sink reactions must be described by two records in the file—one for the source, and one for the sink. Conversely, if a source or sink is not

TABLE 2.10

VolumeSrcSnks.csv

```
Compartment,First SV,Second SV,Rate Constant (/day) wrt First SV
3,1,2,10
3,2,1,-10
3,2,1,90
3,1,2,-90
1,1,1,-0.01
2,1,1,-0.01
3,1,1,-0.01
3,2,2,-0.01
-999
```

mirrored by a complementary sink or source to another state variable, then only the single source or sink is described by a single record.

In the k_s/k_{ds} PRAM-like example, there are both of these types of source and sink reactions: (1) coupled reactions between two modeled state variables where the loss of chemical mass from one state variable represents a gain of mass to another state variable and vice versa, and (2) exogenous sinks and sources (see Chapter 4) of chemical mass representing reactions where, for a sink, the chemical is simply "going away" due to, e.g., biochemical or radioactive decay or, for a source, some type of internal growth process such as cellular division for a microorganism-based state variable. These exogenous sources and sinks are reactions that are not coupled with another state variable—at least not one being modeled.

In Table 2.10, the first four data records represent the first type of source or sink, i.e., conversion of one state variable to another. The first data record (3,1,2,10) describes the source of dissolved chemical in compartment 3 due to desorption of particulate chemical. This source is the term $k_{ds}V_3c_{p3}$ in Equation (2.3), repeated below.

$$V_3 \frac{dc_{d3}}{dt} = \frac{E_{2,3}A_{2,3}}{L_{2,3}}(c_2 - c_{d3}) - kV_3c_{d3} + \mathbf{k_{ds}V_3c_{p3}} - k_sV_3c_{d3} \quad \text{(2.3 repeated)}$$

This source is entered by putting the compartment number in field 1, the two affected state variable numbers in fields 2 and 3, and the rate constant value in field 4. (Our state variable 1 is dissolved chemical and 2 is sorbed chemical in this example.) The First SV and Second SV descriptions of fields 2 and 3 should not be confused with your state variable numbering scheme. "First" does not mean state variable 1, and "Second" does not mean state variable 2. They simply denote a *pair* of state variables.

The combination of the Compartment value (field 1) and the First SV value (field 2) determine the equation that the source or sink term is being added to in the GEM. Because we entered 3 and 1, respectively, as Compartment and First SV, the equation being built is dissolved chemical in compartment 3, i.e., Equation (2.3). (As we describe later, in Chapter 4, the equation number will be the row number of matrix **A**, e.g., row *i*.) The Second SV entry tells the GEM where the source or sink is coming from or going to. (This will the column number of matrix **A**, e.g.. column *j*.)

We have established that we are building Equation (2.3) because of our Compartment and First SV entries. (We established i in $A_{i,j}$.) Considering Equation (2.3), we see three source/sink terms. To which one does our data record refer, that is, what is j in $A_{i,j}$? The Second SV value is 2, so we now know that the data record is specific to term $\mathbf{k_{ds}V_3c_{p3}}$ because that term involves state variable 2 The rate constant value and sign in field 4 must then reflect the value and sign of k_{ds} in that term, i.e., +10.

Let's consider the second data record $(3,2,1,-10)$. The compartment value is 3 and the First SV value is 2, so this record describes a source/sink term in the equation for sorbed chemical in compartment 3, i.e., equation i is (2.4), also repeated below. Which of the three source/sink terms is it, i.e., what is j in $A_{i,j}$? As you will see in Chapter 4, all sink terms appear in the $\mathbf{A}$ matrix in the main diagonal terms, i.e., the $A_{i,i}$ elements. Therefore, when the GEM sees a negative value in field 4, it knows the term is a sink. Once the i (row i) is established from the compartment and First SV fields, that is all the information that is needed, i.e., $j = i$. Our entry of 1 in the Second SV field is purely for our own information (reminding us that this is in fact not only a sink from the perspective of state variable 2 in compartment 3, but also a source relative to state variable 1. Nonetheless, the GEM ignores this field's entry when it sees that the data record describes a sink.

$$V_3 \frac{dc_{p3}}{dt} = k_s V_3 c_{d3} - \mathbf{k_d V_3 c_{p3}} - k V_3 c_{p3} \qquad (2.4 \text{ repeated})$$

Returning to Table 2.10, the third and fourth data records are completely analogous to the first two described above and populate, respectively, the source term $k_s V_3 c_{d3}$ in Equation (2.3) and its complement $-k_s V_3 c_{d3}$ in Equation (2.4). Again, because the fourth data record represents a sink, the Second SV entry is irrelevant for GEM purposes.

The last four data records all describe sinks because field 4 is negative. The first three describe the exogenous decay of dissolved chemical in compartments 1, 2, and 3, respectively. The last describes the exogenous decay of sorbed chemical in compartment 3. None of these is a coupled reaction (among modeled state variables) so each reaction is completely described by a single data record. For example, the seventh data record $(3,1,1,-0.01)$ is relevant to compartment 3 and state variable 1 and populates the sink term $-k V_3 c_{d3}$ in Equation (2.3). Again, the 1 entry in Second SV is for our own information only, and the same modification to the main diagonal element of matrix $\mathbf{A}$ would occur whether this value were 1 or 2.

Although not included in this example, exogenous sources are treated identically to exogenous sinks. Because they are not coupled reactions (among modeled state variables), they are completely described by a single data record. The sign on the field 4 rate constant should be positive because the entries are sources. Unlike exogenous sinks, however, the value of the Second SV field matters. The Second SV field should be identical to the First SV field; otherwise the GEM will think the record is for a coupled reaction and will modify the wrong column of the $\mathbf{A}$ matrix. For example, if the k_s/k_{ds} PRAM-like example that we are currently describing had, instead of exogenous sinks, exogenous sources and the rate constants had the same value as the sinks, but opposite sign, the corresponding VolumeSrcSnks.csv file would be

```
Compartment,First SV,Second SV,Rate Constant (/day) wrt First SV
3,1,2,10
3,2,1,-10
3,2,1,90
3,1,2,-90
1,1,1,0.01
2,1,1,0.01
3,1,1,0.01
3,2,2,0.01
-999
```

In many applications, a user will be modeling only a single-state variable. In these cases, the sources and sinks will be limited to the exogenous-type sources and sinks described above and the entries for the first- and second-state variables will simply be 1 in both fields. Finally, if a problem involves no volume-based sources or sinks, the VolumeSrcSnks.csv file would be entered as

```
Compartment,First SV,Second SV,Rate Constant (/day) wrt First SV
-999
```

AreaSrcSnks.csv—Analogous to the volume-based sources and sinks, area-based sources and sinks involve a transfer rate velocity times a concentration times an interfacial area. The AreaSrcSnks.csv file provides the area-based rate velocities for sources and sinks, and the information needed to properly locate these terms in the **A** matrix. These are the cross-compartment first-order sources and sinks described in Chapter 4. An example of an area-based sink would be settling of a particulate (sorbed) chemical from one compartment to a lower compartment. The corresponding area-based source would be resuspension of the particulate chemical from the lower compartment to the higher one.

None of our PRAM-like examples involves settling or resuspension because we made the simplifying assumption that the water column compartment contained no solids. However, for our purposes here, let's relax that assumption and assume that in our kinetic sorption k_s/k_{ds} approach we have suspended sediments in the water column compartment (compartment 2) that contain sorbed PCB and settle into the sediment compartment (compartment 3). We also assume that we are explicitly modeling dissolved and particulate chemicals as two state variables in the sediments (as we did earlier) and in the other compartments. Under those assumptions, the two mass balance equations for the water column compartment could be expressed as

$$V_2 \frac{dc_{d2}}{dt} = \frac{E_{1,2}A_{1,2}}{L_{1,2}}(c_{d1}-c_{d2}) + \frac{E_{2,3}A_{2,3}}{L_{2,3}}(c_{d3}-c_{d2}) + \frac{E_{2,4}A_{2,4}}{L_{2,4}}(c_{d4}-c_{d2})$$

$$-Q_2c_{d2} - kV_2c_{d2} - k_sV_2c_{d2} + k_dV_3c_{d2}$$

$$V_2 \frac{dc_{p2}}{dt} = -Q_2c_{p2} - kV_2c_{p2} + k_sV_2c_{d2} - k_dV_3c_{p2} - v_sA_{2,3}c_{p2} + v_rA_{2,3}c_{p2}$$

where the new parameters v_s and v_r are, respectively, the settling and resuspension velocities of the solids (L/T). Similarly, the two mass balance equations for the sediment compartment now including particulate settling, resuspension, and "burial" could be expressed as

$$V_3 \frac{dc_{d3}}{dt} = \frac{E_{2,3}A_{2,3}}{L_{2,3}}(c_{d2} - c_{d3}) - kV_3c_{d3} + k_dV_3c_{p3} - k_sV_3c_{d3}$$

$$V_3 \frac{dc_{p3}}{dt} = k_sV_3c_{d3} - k_dV_3c_{p3} - kV_3c_{p3} + v_sA_{2,3}c_{p2} - v_rA_{2,3}c_{p3} - v_bA_{3,?}c_{p3}$$

where v_b is a "burial" velocity, reflecting an exogenous sink of particulate chemical from the modeled sediment compartment due to net sedimentation. (The chemical mass is, of course, not "lost" from the sediments; because the net sedimentation is building up the sediment layer and the modeled sediment compartment is not expanding in volume over time to reflect this expansion—is effectively moving upward—the particulate chemical can be thought of as left behind ["buried"] in the modeled compartment.) The new terms in the above equations involving v_s, v_r, and v_b are the area-based sources and sinks we will now address.

Before doing so, however, please note that the interfacial area term ($A_{3,?}$) in the above equation for the burial sink $-v_bA_{3,?}c_{p3}$ is not resolved given our present compartment configuration for the PRAM-like model. Namely, we do not have a compartment underlying compartment 3 for this burial mass flux. As will be discussed in Chapter 4, this requires introduction of a new "dummy" compartment to accommodate the GEM architecture.

To model this term, the GEM needs to find the interfacial area across which this transfer occurs. To do this, we use the next available sequential integer compartment number (7) and introduce a dummy compartment 7 with the area provided in the appropriate file (Interfaces.csv). (All other files would need updating also.) Accordingly, the burial term would now be $-v_bA_{3,7}c_{p3}$ and the GEM would have access to this area parameter.

Under these new assumptions, our AreaSrcSnks.csv file would be as shown in Table 2.11. This file has a user-determined number of records. The first is a description header record. There is one subsequent record for each area-based source or sink in the model. The -999 record flag is used to signify the end of the file.

In Table 2.11, we show the settling, resuspension, and burial parameters as symbols (v_r) where they would occur in field 4. In practice, of course, one would enter the numerical values and appropriate signs for these parameters in lieu of the symbols.

The rules for populating the AreaSrcSnks.csv file are completely analogous to those discussed above for the VolumeSrcSnks.csv file. In the VolumeSrcSnks.csv file, for a given data record, the compartment was fixed (no cross-state variable transfers are assumed to occur across compartments, but only within a given compartment), and the transfer due to the source or sink was from one state variable to another. For area-based source/sinks for a given data record, the state variable is assumed fixed (again, no cross-state variable transfers are assumed across compartments),

TABLE 2.11

AreaSrcSnks.csv

```
SV,First Compartment, Second Compartment, Rate Const (/day) wrt
First Compartment
2,2,3,-v_s
2,2,3,+v_r
2,3,2,-v_r
2,3,2,+v_s
2,3,7,-v_b
-999
```

and the transfer due to the source or sink is from one compartment to another for a given state variable.

As was the case for volume-based source/sinks, if field 4 has a positive sign, the data record reflects a source with respect to the equation defined by the state variable (field 1) and First Compartment (field 2) combination. This combination determines the correct row i of the **A** matrix. The Second Compartment (field 3) determines the correct column j of the **A** matrix for sources only. When the field 4 entry is negative, the GEM recognizes that the term is a sink, the appropriate entry is on the main diagonal of the **A** matrix ($A_{i,i}$ elements), and the only needed information is the field 1 state variable and the field 2 First Compartment. Any field 3 Second Compartment information is ignored by the GEM.

A caveat is needed to end the discussion of populating the AreaSinks.csv file. Area-based, cross-compartment mass fluxes between a modeled, interior compartment i and an exterior compartment (boundary or dummy) j must be *sinks only* with respect to compartment i, i.e., fluxes *from i to j*, not vice versa. If conditions in compartment j result in area-based fluxes from j to i and j is a dummy compartment, the current capabilities of the GEM have no way to accommodate this mechanism. Indeed, we would argue that if conditions in j are changing dynamically and affecting j's neighbors, j should rightly be included as a modeled interior compartment. If j is a boundary compartment, alternative types of boundary conditions could simulate such a transfer feedback, but the GEM does not presently include them.

Area-based mass fluxes from interior compartment i to boundary or dummy compartment j (*sinks* with respect to i) are treated currently by the GEM simply as exogenous losses from the system without feedback from j to i. If the user enters an area-based transfer from exterior compartment j to interior compartment i (a source with respect to i), an error message will be generated. (Note that if such a source is desired and is known or predictable outside of the system modeled, it could be entered as an external loading in the Loads.csv file.)

LinearKdandTempCoef.csv—LinearKdandTempCoef.csv holds the partition coefficient (K_d) for linear problems considering retardation and the coefficient that allows temperature corrections to kinetic reactions that are not at 20°C. Table 2.12 is the LinearKdandTempCoef.csv file for the K_d version. This file has MaxNSV*NT + 1 records. The first record is a description header. The first field of the data record

TABLE 2.12

LinearKdandTempCoef.csv

```
SV,Compartment,Linear Kd (cu m-water/gm),Temp Coeff
1,1,0,1.024
1,2,0,1.024
1,3,1.80E-06,1.024
1,4,0,1.024
1,5,-999,-999
1,6,-999,-999
```

is the relevant state variable. The second field is the compartment number. The third field is the K_d value and the fourth field is the temperature correction coefficient. Like all of the input files with MaxNSV*NT data records there are MaxNSV blocks of NT records. The blocks must be in sequential order of state variable numbering in field (first field) from 1 to MaxNSV; that is, the first NT records (after the header) relate to state variable 1, the second set of NT records to state variable 2, and so on. Within each block, the records are also sequentially numbered by compartment from 1 to NT in field 2.

The K_d data in this file are used to calculate the retardation coefficient R where, as described in Chapter 4,

$$R = 1 + \frac{BD(K_d)}{\theta}$$

where BD represents solids bulk density and θ is fractional water content (porosity for saturated media). This equation pertains to linear partitioning, i.e., where K_d is a fixed constant. If you are not considering retardation (want $R = 1$) or your partition coefficient is nonlinear, enter $K_d = 0$ in the K_d field. (Data inputs for nonlinear problems including nonlinear K_d functions are described later.) If a state variable–compartment combination is not relevant, specify $K_d = -999$ in that field.

The temperature correction coefficient is ϕ in the $\phi^{(T-20)}$ term discussed in Chapter 4. All first-order source/sink terms in the GEM include this temperature correction. The compartment-specific temperature is in the Compartments.csv file. (In our PRAM-like example, we assumed a constant temperature of 20°C.)

Boundary.csv—Boundary.csv holds the fixed concentration boundary conditions when the boundary type is 2, as specified in field 3 of the Compartments.csv file. Table 2.13 is the Boundary.csv file for the k_s/k_{ds} version. This file has MaxNSV*NT + 1 records. The first record is a description header. The first field of the data record is the relevant state variable. The second field is the compartment number, and the third field is the fixed concentration boundary value (g/m³). The PRAM-like model has compartment 5 as a boundary compartment with a fixed concentration boundary condition of 0.

Like all of the input files with MaxNSV*NT data records, there are MaxNSV blocks of NT records These MaxNSV blocks must be in sequential order of state

TABLE 2.13

Boundary.csv

```
SV,Compartment,Boundary Value (g/m3)
1,1,-999
1,2,-999
1,3,-999
1,4,-999
1,5,0
1,6,-999
2,1,-999
2,2,-999
2,3,-999
2,4,-999
2,5,-999
2,6,-999
```

variable numbering in field (first field) from 1 to MaxNSV; that is, the first NT records (after the header) relate to state variable 1, the second set of NT records to state variable 2, and so on. Within each block, the records are also sequentially numbered by compartment from 1 to NT in field 2. For compartments that are not relevant (any compartment that is not a boundary compartment with a fixed concentration boundary condition type) or state variable–compartment combinations that are not relevant, enter -999 in the boundary value field.

Initial.csv—Initial.csv holds the initial conditions for dynamic simulations. Table 2.14 is the Initial.csv file for the k_s/k_{ds} version. This file has MaxNSV*NT + 1 records. The first record is a description header. The first field of the data record is

TABLE 2.14

Initial.csv

```
SV,Compartment,Initial Condition (g/m3)
1,1,0
1,2,0
1,3,0
1,4,0
1,5,-999
1,6,-999
2,1,-999
2,2,-999
2,3,0
2,4,-999
2,5,-999
2,6,-999
```

the relevant state variable. The second field is the compartment number, and the third field is the initial value (g/m^3). We started the PRAM-like model dynamic simulations with initial conditions of 0 in all compartments.

Like all the input files with MaxNSV*NT data records, there are MaxNSV blocks of NT records. These MaxNSV blocks must be in sequential order of state variable numbering in field (first field) from 1 to MaxNSV; that is, the first NT records (after the header) relate to state variable 1, the second set of NT records to state variable 2, and so on. Within each block, the records are also sequentially numbered by compartment from 1 to NT in field 2. Initial conditions are only relevant to non-boundary or non-dummy compartments, i.e., interior compartments for which concentrations are simulated. For boundary or dummy compartments and/or if a state variable is not relevant in an interior compartment (e.g., state variable 2 in interior compartments 1, 2, and 4 in the PRAM-like example), enter -999 in the initial condition field.

For compartments that are not relevant (any compartment that is not a boundary compartment with a fixed concentration boundary condition type) or state variable–compartment combinations that are not relevant, enter -999 in the boundary value field.

Guess.csv—Guess.csv holds the user's guess for the GEM's solution to the chemical concentrations when a problem is nonlinear. The iterative Newton's method algorithm for nonlinear solutions (see Chapter 5) requires an initial guess to begin its iterations. For dynamic nonlinear problems where the iterative method is applied at each time step, the user's guess is used for the first time step. For all subsequent time steps, the solution from the previous time step is used for the guess. For steady-state problems, the user's guess is used for the (single) solution.

In a later description of the shell functionality (for fofc.exe), we modify the K_d version to include a nonlinear retardation factor. Table 2.15 is the Guess.csv file for that nonlinear K_d version. The guess is 1.0 g/m^3 for the chemical concentrations in each of the four compartments. This file has MaxNSV*NT+ 1 records. The first record is a description header. The first field of the data record is the relevant state variable. The second field is the compartment number, and the third field is the guess (g/m^3). Like all of the input files with MaxNSV*NT data records, there are MaxNSV blocks of NT records These MaxNSV blocks must be in sequential order of state variable numbering in field (first field) from 1 to MaxNSV; that is, the first NT records (after

---

TABLE 2.15

Guess.csv

```
SV,Compartment,Initial Guess (g/m3)
1,1,1
1,2,1
1,3,1
1,4,1
1,5,-999
1,6,-999
```

the header) relate to state variable 1, the second set of NT records to state variable 2, and so on.

Within each block, the records are also sequentially numbered by compartment from 1 to NT in field 2. The guesses are only relevant to non-boundary or non-dummy compartments, i.e., interior compartments for which concentrations are simulated. For boundary and dummy compartments and/or if a state variable is not relevant in an interior compartment, enter -999 in the guess field.

2.2.1.2   Equation Solver Input Files

Although we solved the Chapter 1 F_d version of the PRAM-like model using *Environmental System* mode and made use of the FlowandEMultipliers.csv file, we will not illustrate how to solve it in *Equation Solver* mode. We illustrate the static *Equation Solver* mode input files and will base these input files on the dynamic form of that example. Each time step of the dynamic solution involves four equations in four unknowns (dissolved PCB concentrations in each compartment).

AEqnSolver.csv—AEqnSolver.csv is the input file that contains the linear coefficients in the **A** matrix. Table 2.16 is the AEqnSolver.csv file for the F_d version. This file has NEQN records, each with NEQN fields, where NEQN is the number of equations. For example, the 1,1 element (–19,200) is the coefficient of c_1 from Equation (1.10a) in Chapter 1, while the 1,2 element (18,000) is the coefficient of c_2 in that equation.

BEqnSolver.csv—BEqnSolver.csv is the input file that contains the loads and boundary conditions. Table 2.17 is the BEqnSolver.csv file for the F_d version. This file has NEQN records, each with a single field. Each field can contain loads and/or boundary conditions.

*A **word of caution***—For steady-state problems, the **b** vector (i.e., BEqnSolver.csv file) containing the load and boundary conditions has negative entries because the **b** vector has been moved to the right side of the algebraic matrix equation $A\underline{c} = -\underline{b}$. For dynamic problems, the elements of the **b** vector are positive because the problem being solved is now the differential equation

$$\frac{d(\mathbf{RV}\underline{c})}{dt} = \mathbf{A}\underline{c} + \underline{b}.$$

When one is switching back and forth between steady-state and dynamic modes, it is easy to forget this sign difference.

TABLE 2.16

AEqnSolver.csv

```
-19200,18000,0,0
18000,-338800,12000,180000
0,120000,-12018,0
0,180000,0,-280000
```

TABLE 2.17	TABLE 2.18
BEqnSolver.csv	**RVEqnSolver.csv**
1000	120000
0	1080000
0	1800
0	1080000

TABLE 2.19	TABLE 2.20
InitialEqnSolver.csv	**GuessEqnSolver.csv**
0	1.0
0	1.0
0	1.0
0	1.0

RVEqnSolver.csv—RVEqnSolver.csv is the input file that contains the retardation (R) and volume (V) data. Each entry is R*V. Table 2.18 is the RVEqnSolver.csv file for the F_d version. (The retardation factor is 1.0 for each equation for this example.) This file has NEQN records, each with a single field.

InitialEqnSolver.csv—InitialEqnSolver.csv is the input file that contains the initial conditions. Table 2.19 is the InitialEqnSolver.csv file for the F_d version under the dynamic scenario. This file has NEQN records, each with a single field.

GuessEqnSolver.csv—GuessEqnSolver.csv holds the user's guess for the GEM's solution to the chemical concentrations when a problem is nonlinear. The iterative Newton's method algorithm for nonlinear solutions requires an initial guess to begin its iterations. For dynamic nonlinear problems where the iterative method is applied at each time step, the user's guess is used for the first time step. For all subsequent time steps, the solution from the previous time step is used for the guess. For steady-state problems, the user's guess is used for the (single) solution.

Although we did not provide a nonlinear scenario for the F_d-based PRAM-like model in Chapter 1, Table 2.20 illustrates what the GuessEqnSolver.csv file might look like for a nonlinear variation on that problem. The guess is 1.0 g/m^3 for each variable. This file has NEQN records, each with a single field.

2.2.2 DYNAMIC INPUT FILES USING SHELL PROCEDURE

When the GEM runs, it receives inputs from user-supplied data files and by making internal calculations. When those inputs need to vary dynamically and/or the GEM's internal calculations need assistance from the user, this external help is provided by the use of a shell procedure that enables the GEM code to run an external program, read the program's outputs, and proceed as illustrated in Figure 2.1. See Box 2.2 for an alternative approach to dynamic changes that does not require a shell procedure.

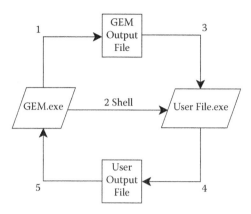

FIGURE 2.1   GEM shell procedure.

**BOX 2.2   ALTERNATIVE APPROACH TO
TIME-VARYING PARAMETERS**

Dynamically changing parameters can also be supplied to the GEM by building external software that not only provides the GEM input files but calls the GEM serially each time the parameter files change during a dynamic simulation. Each serial call to the GEM results in a "complete" GEM run as far as the system is concerned. The GEM sees the set of input files and the number of requested time steps, makes its run, generates output files, and stops. The external software would then read those output files (concentrations), use those ending concentrations as the initial conditions for the next GEM run, build the new parameter files, update internal time-stepping data, and call the GEM again.

This is the approach used by RTI, International with the GEM to develop a new compartment model of the fate and transport of toxic chemicals contained in sewage biosolids when those biosolids are land-applied to farm fields for nutritional augmentation. The model will be used by the U.S. EPA to assess human health risks from biosolids use.

For example, suppose that flows need to vary dynamically during a GEM run. At each time step, the procedure would: (1) write an output file containing the current simulation time, (2) shell to run the external user-supplied executable 'userfile'.exe file, (3) 'userfile'.exe would read the current simulation time, (4) calculate or obtain from some external source the appropriate flows and write them to its output file, which is the GEM input file; and (5) read updated flows from the new file and continue.

In addition to updating dynamic inputs, this shell procedure is useful for another purpose. For nonlinear problems, the GEM cannot anticipate the wide variety of functional forms that the nonlinearities might take on. The shell procedure allows a user to specify that functionality to the GEM. This has nothing to do per se with updating dynamic data. It is a case of providing external information to the GEM about functional relationships among variables for steady-state and dynamic problems.

The naming convention for the files involved in the shell procedure is the following. The user-provided executable file must be named <input file prefix> + .exe. For example, the executable that will read the GEM-created output file and update the input file Flows.csv must be named Flows.exe. The executable that will update the input file AEqnSolver.csv must be named AEqnSolver.exe. The output file that the GEM writes prior to shelling to the user's executable is named Shell + <prefix of executable file> + .csv, e.g. ShellFlows.csv or ShellAEqnSolver.csv. This file will contain for dynamic problems the time (days) at the current time step as the first record and the current values of the NEQN state variable concentrations as the next NEQN record.

Each record has a single field. No descriptive headers are written. For steady-state problems, no time record is written, i.e., the NEQN records are the NEQN state variable concentrations. (For steady-state nonlinear problems, these concentrations will change iteratively during the Newton algorithm. For linear dynamic problems, the time and concentrations will change at each time step. For nonlinear dynamic problems the time will change at each time step and the concentrations will change at each time step and Newton iteration.) Finally, the file that the user's executable writes must have the same name as shown in Table 2.1. For *Environmental System* or *Equation Solver* mode, when these files are dynamically written, they must have the identical structure (including a descriptive header record if needed) as described in the static file descriptions or below.

Table 2.1 presented all the files that can be written dynamically including (1) files that can be written statically or dynamically, and (2) files that are written only dynamically. This latter category includes Fofc.csv, Dfdc.csv, Rofc.csv, and dRdc. csv. These four files have the same names and formats for both *Environmental System* and *Equation Solver* modes and would be created analogously under either mode. These are the only dynamic files that we describe and illustrate. Files in the former category (dynamically or statically written) would be dynamically written using procedures analogous to those we describe below.

For those users who acquired the GEM source code, no external shelling is necessary. Analogous code that would be in the external executables may be inserted integrally into the appropriate GEM subroutine. For those users desiring to develop external code for use in the shell procedure, but who do not have programming experience, a brief overview of the current programming environment, and our thoughts on it, are provided in Box 2.3.

fofc.exe—fofc.exe is a user-supplied executable code that populates the fofc.csv dynamic file. As described in Chapter 4, the most general GEM mass balance equation (written in matrix form) is

$$\frac{d(\mathbf{R}(\underline{c})\mathbf{V}\underline{c})}{dt} = \mathbf{A}\underline{c} + \underline{f}(\underline{c}) + \underline{b}$$

where $\underline{f}(\underline{c})$ represents generalized (user-specified), nonlinear functionality in the source/sink terms. Fofc.exe informs the GEM of the functionality for the term. This is not a data file (like the .csv files) but a compiled computer program.

BOX 2.3   OOPS! THE NON-PROFESSIONAL PROGRAMMER

The GEM has been written in Visual Basic (VB 6.0) and is in development in Java. Both have advantages and disadvantages for an experienced programmer. However, we realize that many readers are not experienced programmers and now that we're asking you to write your own programs for the shell functionality, a brief overview of the current programming environment is offered from our own admittedly biased perspective.

Once upon a time, there was a golden era when the scientist or engineer could single-handedly investigate scientific theory and write a computer program to implement that theory. Writing a computer program in FORTRAN or earlier versions of the BASIC language was a relatively easy task and the ability to do so was simply another tool in the technical toolbox—not much different in principle from using a calculator to perform your job. This pleasant situation existed because the earlier languages were straightforward (procedural) and conversational (no arcane commands), and one did not have to be a software engineer or computer scientist to use them effectively.

No more. That situation changed with the introduction of object-oriented programming (OOP) in the 1990s. OOP undoubtedly has made possible highly flexible software that can perform certain tasks (touch screens, web programming, other graphical user interfaces [GUIs]) that were not possible with the old software tools. The cost of these benefits, however, has been the steady decline in accessibility to software development for non-professional programmers like the target audience for this book. As an example, compare the code needed to open a file and read data from it under earlier versions of BASIC (VB 6, circa 2005) and the current version (VB.net) and its OOP-based commands.

VB 6.0

```
Open "Test.txt" for Input as #1
        Input #1, x
Close #1
```

VB.net (from Mansfield, 2005)

```
Dim strFileName as String = "C:\Test.txt"
Dim objFileName as FileStream = New FileStream(strFileName,
    FileMode.Open,FileAccess.Read, FileShare.Read)
Dim objFileRead As StreamReader = New StreamReader(objFilename)
While (objFileRead.Peek() > -1)
        textbox1.Text &= objFileRead.Readline()
End While
objFileRead.Close()
objFilename.Close()
TextBox1.Select(0,0)
```

This abandonment of a language accessible to "the rest of us" (who, by the way, constitute the vast majority of programmers) was met with outrage from the non-professional programmer community, and many abandoned programming—or at least Microsoft's BASIC—with obvious consequences for Microsoft's sales. In time, Microsoft saw its error and reached out to the non-professional community with VB Express. VB Express is not quite a return to the simple elegance of VB 6, but it is an attempt. Our target audience for this book—the prospective modeler—may want to consider VB Express. Best of all, it's free. (http://www.microsoft.com/express/Downloads/#2008-Visual-Basic).

Many OOP programmers ridiculed the earlier BASIC languages as "toys," not suitable for serious programming. That is simply not true. For number-crunching, which is the core activity of scientific programming, the procedural BASIC languages were and are as good as any and better than many. Add to that the huge advantage of not needing a team of software engineers to build your application—with attendant costs and "tower of Babel" communication problems—and that is a hard combination to beat.

For example, suppose for our k_s/k_{ds} version that the PCB decay terms are second-order sinks instead of first-order terms. First-order source/sink terms are linear functions of concentration whereas higher-order source/sinks are nonlinear. With second-order decay terms, the five compartmental mass balance equations are (recall that the air compartment does not have decay) listed below. We ordered them as they will be numbered internally in the GEM, as explained below.

Equation 1, state variable 1 (dissolved c) in compartment 1 (ship compartment):

$$V_1 \frac{dc_1}{dt} = W + \frac{E_{1,2}A_{1,2}}{L_{1,2}}(c_2 - c_1) - kV_1c_1^2$$

Equation 2, state variable 1 (dissolved c) in compartment 2 (water column compartment):

$$V_2 \frac{dc_2}{dt} = \frac{E_{1,2}A_{1,2}}{L_{1,2}}(c_1 - c_2) + \frac{E_{2,3}A_{2,3}}{L_{2,3}}(c_{d3} - c_2) + \frac{E_{2,4}A_{2,4}}{L_{2,4}}(c_4 - c_2) - Q_2c_2 - kV_2c_2^2$$

Equation 3, state variable 1 (dissolved c) in compartment 3 (sediment compartment):

$$V_3 \frac{dc_{d3}}{dt} = \frac{E_{2,3}A_{2,3}}{L_{2,3}}(c_2 - c_{d3}) - kV_3c_{d3} + k_dV_3c_{p3} - k_sV_3c_{d3}^2$$

Equation 4, state variable 1 (dissolved c) in compartment 4 (air compartment):

$$V_4 \frac{dc_4}{dt} = \frac{E_{2,4} A_{2,4}}{L_{2,4}} (c_2 - c_4) - Q_4 c_4$$

Equation 5, state variable 2 (particulate c, c_p) in compartment 3 (sediment compartment):

$$V_3 \frac{dc_{p3}}{dt} = k_s V_3 c_{d3} - k_d V_3 c_{p3} - k V_3 c_{p3}^2$$

Under this assumption of nonlinear source/sink terms, the nonlinear $\underline{f(c)}$ vector would be (and the VolumeSrcSnks.csv file would contain no source/sink data, i.e., just the header record and the "-999" end of file flag)

$$\underline{f}(\underline{c}) = \begin{bmatrix} -k V_1 c_1^2 \\ -k V_2 c_2^2 \\ -k V_3 c_{3d}^2 \\ 0 \\ -k V_3 c_{3p}^2 \end{bmatrix}$$

The fofc.exe program that the user supplies will provide the values in the $\underline{f(c)}$ vector to the GEM. The GEM will first write the Shellfofc.csv with the concentrations and the current time step for dynamic problems. It will then shell out and fire the user's fofc.exe program. The fofc.exe program reads the concentrations and current time (if a dynamic problem) from the Shellfofc.csv file, calculates the elements of the $\underline{f(c)}$ vector, and writes those data to the GEM input file, fofc.csv. (The fofc.csv file contains data, i.e., nonlinear functions evaluated at the current time step iteration, not the functions themselves.) The GEM reads fofc.csv and continues its solution.

Previously, in preparing the GEM's .csv input data files, the user has only been asked to populate the data files by reference to compartment number and/or state variable number. It has not been necessary to understand how the GEM builds its internal system of equations from these data. For practical reasons, it is now necessary in developing fofc.exe that the user understand this procedure and provide the elements of the $\underline{f(c)}$ vector in that format.

The $\underline{f(c)}$ vector has the same number of elements as the number of equations being solved. This number is NEQN, and is the number of state variables relevant to each interior (non-boundary or dummy) compartment summed over all interior compartments. The GEM calculates NEQN as

$$\text{NEQN} = \sum_{i=1}^{MaxNSV} \sum_{j=1}^{NT} I_{i,j}$$

where

$$I_{i,j} = \begin{bmatrix} 1, \text{if SV i is relevant to compartment j} \\ 0, \text{otherwise} \end{bmatrix}$$

For example, in our k_s/k_{ds} version, we have four interior compartments and two boundary/dummy compartments so NT = 6. In three of these (compartments 1, 2, and 4), there is only one relevant state variable (dissolved PCB). The sediment compartment (compartment 3) contains two relevant state variables so MaxNSV = 2. Thus, the above sum would yield NEQN = 5 as

+1 (SV 1, interior compartment 1)
+1 (SV 1, interior compartment 2)
+1 (SV 1, interior compartment 3)
+1 (SV 1, interior compartment 4)
+0 (SV 1, boundary compartment 5)
+0 (SV 1, dummy compartment 6)
+0 (SV 2, interior compartment 1)
+0 (SV 2, interior compartment 2)
+1 (SV 2, interior compartment 3)
+0 (SV 2, interior compartment 4)
+0 (SV 2, boundary compartment 5)
+0 (SV 2, dummy compartment 6)
  5 (total)

If this all seems somewhat familiar, it is, because the SVCompMap.csv file is ordered identically. Each record of the SVCompMap.csv file corresponds to a (sequentially numbered) state variable. Thus, the MaxNSV records comprise the first of the double sums in the above equation. Within each record, the first field is the state variable index number. After the first field, however, summing across the remaining NT fields corresponds to the second sum in the above equation.

This scheme is also how the SV–compartment combinations are ordered in the NEQN system of equations. That is, in the above sum, the equations are ordered as the 1s are encountered. For example, the equation corresponding to SV 2 in compartment 3 is 5, because it is the fifth 1 encountered.

As an example of fofc.exe, the following Visual Basic code would populate the fofc. csv file for the k_s/k_{ds} version. The current time is read, but is not needed in this example.[*]

---

[*] For example, if the volumes were changing dynamically, the current time would be needed to correctly update them.

```
Private Sub Form_Load()
Dim I As Integer
Dim NEQN As Integer
NEQN = 5
Dim Time As Single
ReDim c(NEQN) As Single
ReDim fofc(NEQN) As Single
Dim V1 As Single
Dim V2 As Single
Dim V3 As Single
Dim V4 As Single
V1 = 120000
V2 = 1080000
V3 = 1800
V4 = 1080000
Dim k As Single
k = 0.01

Open "Shellfofc.csv" For Input As #1
  Input #1, Time
  For I = 1 To NEQN
    Input #1, c(I)
  Next I
Close #1

fofc(1) = -k * V1 * c(1) ^ 2
fofc(2) = -k * V2 * c(2) ^ 2
fofc(3) = -k * V3 * c(3) ^ 2
fofc(4) = 0
fofc(5) = -k * V3 * c(5) ^ 2

Open "fofc.csv" For Output As #1
  For I = 1 To NEQN
    Print #1, fofc(I)
  Next I
Close #1
End
End Sub
```

The reader may note that we entered the compartment volumes after they were already entered in Compartments.csv. The user could elect to read them into the fofc.exe program from Compartments.csv. We took the redundant but easy way out for this example.

We ran the GEM on the PRAM-like example using the second-order nonlinear decay term for $\underline{\mathbf{f}(\mathbf{c})}$. All other parameters and inputs remained as in the kinetic sorption/desorption version of the model. Recall from our earlier examples that the concentrations in the four compartments were all less than 1.0 g/m^3. Therefore, if our new second-order decay term is applying the −0.01/day rate constant to the concentrations squared, you would expect the concentrations from the second-order assumption to be greater than under the first-order assumption. At steady state (running the GEM under steady-state mode or dynamic mode out to steady state), the new concentrations were

c_1 (ship) = 6.86×10^{-2} g/m^3
c_2 (water compartment) = 1.34×10^{-2} g/m^3
c_3 (dissolved in sediment) = 1.34×10^{-2} g/m^3
c_4 (air) = 8.59×10^{-3} g/m^3
c_5 (particulate in sediment) = 0.120 g/m^3

Indeed they are somewhat greater than the first-order results.

Dfdc.exe—Dfdc.exe is a user-supplied executable code that populates the Dfdc.csv dynamic file. This executable file is required for nonlinear problems if you have an $\underline{f}(\underline{c})$ function *and* are using analytical derivatives in the Jacobian matrix. As discussed in Chapter 5, the Jacobian matrix is the NEQN × NEQN matrix of partial derivatives of the elements of the implicit function $\underline{g}(\underline{c})$ with respect to the elements of the chemical concentration vector $\underline{c}$. The Newton algorithm finds the "roots" (zeros) of $\underline{g}(\underline{c})$ for a steady-state problem,

$$\underline{g}(\underline{c}) = \mathbf{A}\underline{c} + \underline{f}(\underline{c}) + \underline{b} = \underline{0}$$

where
  $\underline{c}$ = NEQN × 1 vector of unknown concentrations to be determined
  $\mathbf{A}$ = NEQN × NEQN matrix of linear coefficients of c
  $\underline{f}(\underline{c})$ = NEQN × 1 vector of nonlinear source/sink terms
  $\underline{b}$ = NEQN × 1 vector of external loadings
These terms are explained in detail in later chapters of this book. The i,jth element of the Jacobian matrix for steady-state problems is

$$\frac{\partial g_i}{\partial c_j} = A_{i,j} + \frac{\partial f_i}{\partial c_j}$$

where $A_{i,j}$ denotes the i,jth element of the $\mathbf{A}$ matrix and $\partial f_i/\partial c_j$ is the partial derivative of the ith row of the nonlinear source/sink vector $\underline{f}(\underline{c})$ with respect to c_j. The dfdc.exe file will write an output file (dfdc.csv) that contains the NEQN × NEQN matrix of $\partial f_i/\partial c_j$ for i and j = 1, …, NEQN *evaluated* at the current time and Newton iteration. Again, these outputs are *data*, not the analytical derivatives.

The GEM provides as input to dfdc.exe in Shelldfdc.csv the current time and current values of c_i, i = 1, …, NEQN. For steady-state problems, the time is not written. Given the dfdc.exe-supplied values of $\partial f_i/\partial c_j$, the GEM then constructs the complete Jacobian matrix by adding those data to the $A_{i,j}$ values it has already constructed internally from the coefficients of the linear terms of the problem and $\partial R_i/\partial c_j$ elements if relevant (see Rofc.exe description).

Again, it is assumed that the user understands the structure of the NEQN equations (the ordering of state variables and interior compartments), as described in the fofc.exe discussion above. Thus, when the dfdc.exe program writes $\partial f_i/\partial c_j$, that element will correspond to the ith row and the jth column of the Jacobian, and the user must understand which state variable or compartment the ith row represents and which state variable or compartment the jth column represents. For example, for the k_s/k_{ds} version with nonlinear sinks,

$$\underline{\mathbf{f}}(\mathbf{c}) = \begin{bmatrix} -kV_1c_1^2 \\ -kV_2c_2^2 \\ -kV_3c_{3d}^2 \\ 0 \\ -kV_3c_{3p}^2 \end{bmatrix}$$

the matrix of partial derivatives is

$$\begin{bmatrix} -2kV_1c_1 & 0 & 0 & 0 & 0 \\ 0 & -2kV_2c_2 & 0 & 0 & 0 \\ 0 & 0 & -2kV_3c_{3d} & 0 & 0 \\ 0 & 0 & 0 & 0 & 0 \\ 0 & 0 & 0 & 0 & -2kV_3c_{3p} \end{bmatrix}$$

and dfdc.exe would evaluate each of these elements at the c values associated with the current time step and Newton iteration (for steady-state problems, the c values vary only by Newton iteration) and return those evaluated results to dfdc.csv. A Visual Basic program that will perform that task for this $\underline{\mathbf{f}}(\underline{\mathbf{c}})$ is shown below.

```
Dim I As Integer
Dim J As Integer
Dim NEQN As Integer
NEQN = 5
Dim Time As Single
ReDim c(NEQN) As Single
ReDim dfdc(NEQN, NEQN) As Single
Dim V1 As Single
Dim V2 As Single
Dim V3 As Single
Dim V4 As Single
V1 = 120000
V2 = 1080000
V3 = 1800
V4 = 1080000
```

```
Dim k As Single
k = 0.01

Open "Shelldfdc.csv" For Input As #1
  Input #1, Time
  For I = 1 To NEQN
    Input #1, c(I)
  Next I
Close #1

For I = 1 To NEQN
  For J = 1 To NEQN
    dfdc(I, J) = 0 'initialize
  Next J
Next I

dfdc(1, 1) = -2 * k * V1 * c(1)
dfdc(2, 2) = -2 * k * V2 * c(2)
dfdc(3, 3) = -2 * k * V3 * c(3)
dfdc(5, 5) = -2 * k * V3 * c(5)

Open "dfdc.csv" For Output As #1
  For I = 1 To NEQN
    For J = 1 To NEQN
      If J < NEQN Then
        Print #1, dfdc(I, J); ",";
      Else 'don't append comma to end of row
        Print #1, dfdc(I, J)
      End If
    Next J
    Next I
Close #1

End
End Sub
```

When this dfdc.exe is executed for *c* values of 1.0, the dfdc.csv file written is

```
-2400 , 0 , 0 , 0 , 0
 0 ,-21600 , 0 , 0 , 0
 0 , 0 ,-36 , 0 , 0
 0 , 0 , 0 , 0 , 0
 0 , 0 , 0 , 0 ,-36
```

For our above example, the $\mathbf{f(c)}$ function resulted in non-zero derivatives only on the main diagonal of the matrix. It should be noted, however, that it is possible to have non-zero elements *anywhere* in the dfdc.csv file. For example, suppose our k_s/k_{ds} version sorption/desorption processes were also second-order. $\mathbf{f(c)}$ would then be

$$\mathbf{\underline{f}(\underline{c})} = \begin{bmatrix} -kV_1c_1^2 \\ -kV_2c_2^2 \\ -kV_3c_{d3}^2 - k_sV_3c_{d3}^2 + k_{ds}V_3c_{p3}^2 \\ 0 \\ -kV_3c_{3p}^2 + k_sV_3c_{d3}^2 - k_{ds}V_3c_{p3}^2 \end{bmatrix}$$

and the matrix of partial derivatives would be

$$\begin{bmatrix} -2kV_1c_1 & 0 & 0 & 0 & 0 \\ 0 & -2kV_2c_2 & 0 & 0 & 0 \\ 0 & 0 & (-2kV_3c_{3d} - 2k_sV_3c_{3d}) & 0 & 2k_{ds}V_3c_{3p} \\ 0 & 0 & 0 & 0 & 0 \\ 0 & 0 & 2k_sV_3c_{3p} & 0 & (-2kV_3c_{3p} - 2k_sV_3c_{3p}) \end{bmatrix}$$

The dfdc.exe would write these data accordingly.

Rofc.exe—Rofc.exe is a user-supplied executable code that populates the Rofc. csv dynamic file. For steady-state problems, retardation (linear or nonlinear) is not relevant. Considering again the generalized GEM mass balance matrix equation

$$\frac{d(\mathbf{R(\underline{c})V\underline{c}})}{dt} = \mathbf{A\underline{c}} + \mathbf{\underline{f}(\underline{c})} + \mathbf{\underline{b}}$$

$\mathbf{R(\underline{c})}$ is a term that represents generalized (user-specified), nonlinear functionality in the retardation term (see Chapter 4). Linear retardation is included via the LinearKdandTempCoef.csv file previously described. Rofc.exe informs the GEM of the functionality for the nonlinear term. This is not a data file (like the .csv files) but a compiled computer program.

$\mathbf{R(\underline{c})}$ is shown as an NEQN × NEQN matrix in the generalized mass balance equation. In fact, however, only the main diagonal elements of $\mathbf{R(\underline{c})}$ $(i,i$ elements) contain the retardation information, the off-diagonal elements $(i,j$ for $i \neq j)$ are 0 everywhere. It is necessary to specify $\mathbf{R(\underline{c})}$ as a matrix instead of a vector so that the matrix multiplication is conformable. The same is true for the matrix $\mathbf{V}$ that contains only the compartment volumes on the main diagonal and zeroes elsewhere. In our K_d version, the dissolved PCB in the sediment compartment was modeled as

$$R_3V_3 \frac{dc_3}{dt} = \frac{E_{2,3}A_{2,3}}{L_{2,3}}(c_2 - c_3) - kV_3c_3$$

where, for linear retardation,

$$R_3 = 1 + \frac{BD_3 K_d}{\theta_3}$$

Suppose, however, that we wanted to use a nonlinear partition coefficient. For the Freundlich-type isotherm, the retardation factor would be expressed as (see Chapter 4)

$$R_3 = 1 + \frac{BD_3 K_d c_{3d}^{N-1}}{\theta_3}$$

where N is an additional retardation parameter, making the term nonlinear (at least for $N \neq 1.0$). For this modified K_d version of the PRAM-like model with $N = 1.5$ and (as before) $K_d = 1.8 \times 10^{-6}$ m³/g, $BD_3 = 2.5 \times 10^6$ g/m³, and $\theta = 0.5$, the **R(c)** matrix would be (recall that this version of the PRAM-like model has one state variable and four equations)

$$\begin{bmatrix} 1 & 0 & 0 & 0 \\ 0 & 1 & 0 & 0 \\ 0 & 0 & (1+9.0c_{3d}^{0.5}) & 0 \\ 0 & 0 & 0 & 1 \end{bmatrix}$$

Note that the 1,1 and 2,2, and 4,4 elements are unity, reflecting the fact that no retardation (linear or otherwise) is being considered for those equations. Note also that if even only one of your NEQN equations involves nonlinear retardation, the GEM will read the retardation coefficients for *all* equations from the **R(c)** matrix.

dRdc.exe would evaluate the main diagonal elements at the c values associated with the current time step and Newton iteration and return those evaluated results to Rofc.csv. Only main diagonal element evaluations are reported, again because all evaluations elsewhere are 0. A Visual Basic program that will perform that task for this **R(c)** is shown below.

```
Private Sub Form_Load()
Dim I As Integer
Dim J As Integer
Dim NEQN As Integer
NEQN = 4
Dim Time As Single
ReDim c(NEQN) As Single
ReDim Rofc(NEQN) As Single
Dim kds As Single
kds = 0.0000018
Dim BD As Single
```

```
BD = 2500000#
Dim watercontent As Single
watercontent = 0.5
Dim N As Single
N = 1.5

Open "ShellRofc.csv" For Input As #1
  Input #1, Time
  For I = 1 To NEQN
    Input #1, c(I)
  Next I
Close #1

For I = 1 To NEQN
  Rofc(I) = 1 'initialize
Next I
Rofc(3) = 1 + (BD * kds * c(3) ^ (N - 1))/watercontent

Open "Rofc.csv" For Output As #1
  For I = 1 To NEQN
    Print #1, Rofc(I)
  Next I
Close #1

End
End Sub
```

When this Rofc.exe is executed for c values of 1.0, the Rofc.csv file that is written is

```
0
0
1
0
```

dRdc.exe—dRdc.exe is a user-supplied executable code that populates the dRdc.csv dynamic file. This executable file is required for nonlinear problems if you have a nonlinear $\mathbf{R(\underline{c})}$ function *and* are using analytical derivatives in the Jacobian matrix.

The dRdc.exe file will write an output file (dRdc.csv) that contains the NEQN main diagonal elements matrix of $\partial R_i/\partial c_j$ for i and $j = 1, ..., $ NEQN *evaluated* at the current time and Newton iteration, i.e., these outputs are again *data*, not the analytical derivatives. It is understood that the off-diagonal elements $(i \neq j)$ are 0. The GEM provides as input to dRdc.exe in ShelldRdc.csv the current time (cumulative time in days at current time step) and current values of c_i, $i = 1, ..., $ NEQN. Given the dRdc.exe-supplied values of $\partial R_i/\partial c_i$, the GEM then constructs the complete Jacobian matrix by adding those data to the $A_{i,j}$ values that it already constructed internally from the coefficients of the linear terms of the problem and $\partial f_i/\partial c_j$ elements (see fofc.exe description) if relevant.

Again, it is assumed that the user understands the structure of the NEQN equations (ordering of state variables and interior compartments), as described previously. Thus, when the dRdc.exe program writes $\partial R_i / \partial c_j$, that element will correspond to the ith row and the ith column of the Jacobian, and the user must understand which state variable or compartment the ith row represents and which state variable or compartment the ith column represents.

For the above K_d version using the Freundlich isotherm in sediment compartment 3, the analytical derivative of the compartment 3 R term is

$$\frac{\partial R_3}{\partial c_{3d}} = \frac{BD_3 K_d (N-1) c_{3d}^{N-2}}{\theta_3}$$

For $N = 1.5$, and the other parameter values as above, the complete matrix of $\partial R_i / \partial c_j$ for i and $j = 1, ..., $ NEQN is

$$\begin{bmatrix} 0 & 0 & 0 & 0 \\ 0 & 0 & 0 & 0 \\ 0 & 0 & (4.5 c_{3d}^{-0.5}) & 0 \\ 0 & 0 & 0 & 0 \end{bmatrix}$$

and dRdc.exe would evaluate the main diagonal elements at the c values associated with the current time step and Newton iteration and return those evaluated results to dRdc.csv. Only main diagonal element evaluations are reported, again because all evaluations elsewhere are 0. A Visual Basic program that will perform that task for this $\mathbf{R(c)}$ is shown below.

```
Private Sub Form_Load()
Dim I As Integer
Dim J As Integer
Dim NEQN As Integer
NEQN = 4
Dim Time As Single
ReDim c(NEQN) As Single
ReDim dRdc(NEQN) As Single
Dim kds As Single
kds = 0.0000018
Dim BD As Single
BD = 2500000#
Dim watercontent As Single
watercontent = 0.5
Dim N As Single
N = 1.5
```

```
Open "ShelldRdc.csv" For Input As #1
  Input #1, Time
  For I = 1 To NEQN
    Input #1, c(I)
  Next I
Close #1

For I = 1 To NEQN
  dRdc(I) = 0 'initialize
Next I
dRdc(3) = (BD * k_ds/watercontent) * (N - 1) * c(3) ^ (N - 2)

Open "dRdc.csv" For Output As #1
  For I = 1 To NEQN
    Print #1, dRdc(I)
  Next I
Close #1

End
End Sub
```

When this dRdc.exe is executed for c values of 1.0, the dRdc.csv file that is written is

```
0
0
4.5
0
```

2.2.3   OTHER CONTROL.CSV INPUTS

In this section, we discuss inputs in the Control.csv file that did not receive adequate or any discussion previously.

If EnvS, enter around-compartments flow balance tolerance (%)—Each time the **A** matrix is loaded (at the beginning of the run, and again anytime the shell functionality is used to update any parameters in the A matrix) an internal flow balance is performed around all interior compartments. Cumulative flows entering all compartments are expected to be balanced by cumulative flows leaving the compartments. If the absolute value of the difference in cumulative inflows and outflows is greater than the user-specified flow balance tolerance (a percentage of cumulative inflows), an error message is written to the Error.dng file and GEM execution is terminated. The flow balance check is *only* for each snapshot in time when the A matrix is loaded. To the extent that flows are varying over time—as provided by the user—temporal flow balance errors are not detected and are the user's responsibility. In addition, if you are modeling a system such as a reservoir involving flow storage and incoming and outgoing flows do not balance but changes in compartment volume occur, you can set the flow balance tolerance to a high percentage to override this feature.

If no analytical derivatives, enter finite difference multiplier for derivative estimation—In the absence of analytical derivatives for the Jacobian matrix, the GEM includes an option to estimate the Jacobian elements numerically. A central differencing method is used in which the i,jth element of the Jacobian matrix is estimated as

$$J_{i,j} = \frac{\partial g_i}{\partial c_j} \approx \frac{g_i(c_i + \varepsilon c_i) - g_i(c_i - \varepsilon c_i)}{2\varepsilon c_i}$$

where

$g_i = i$th element of the NEQN × 1 $\mathbf{g(\underline{c})}$ implicit function
$c_j = j$th variable (concentration)
ε = finite difference multiplier

We have had good success with the examples in this book using $\varepsilon = 0.1$. With this value, the finite difference approximation evaluates the g_i function at ±10% of the current value of c_j.

Warning: Using the numerical approximation option for estimating the Jacobian involves substantially more calculation than use of analytical derivatives. (For each of the NEQN^2 elements of the Jacobian, g_i must be evaluated twice. Each evaluation involves running your fofc.exe and/or Rofc.exe shelled executable programs.) You will not notice this additional computation for small problems, but for problems with relatively large NEQN values, the runtime will be substantially slowed.

If nonlinear, will zero or negative concentrations present numerical problems during Newton method iterations? (Y,N)—For many nonlinear problems, trial values of the unknown concentrations that become 0 or negative during the iterative process will cause mathematical errors. For example, in the Freundlich partitioning isotherm, the concentration is raised to an exponent that may be negative. If a trial value of the concentration becomes 0, then 0 to a negative exponent results in an error. If this is a potential problem, enter Y and the GEM will avoid these intermediate results. (If it is not a problem, it is often advantageous to visit 0 or negative values during the Newton method iteration [even if those values are nonsensical as final solutions] because that may be the shortest route to the solution.)

If nonlinear, enter (1) MaxNewtonIteration, (2) MaxSamplingIteration, (3) NumSamples—The GEM's quasi-Newton algorithm involves two embedded iterative loops. The inner loop is the quasi-Newton algorithm that performs the "Newton step" to derive new trial values beginning from an initial guess. The first time the inner loop is activated, the guess is that provided by the user in the Guess. csv (or GuessEqnSolver.csv) file. The inner loop runs until either convergence or MaxNewtonIteration is reached. If MaxNewtonIteration is reached, the outer loop is then activated to generate a new, randomly-generated guess. NumSamples is the number of random samples that are drawn. For each of the NumSamples random samples, the corresponding f-value is computed and the random sample resulting in the minimum f-value is then used as the new guess. The inner quasi-Newton iteration loop is then reactivated to operate from this new guess. This inner/outer loop process is repeated until MaxSamplingIteration is reached or convergence is attained. If MaxSamplingIteration is reached before convergence, an error message is generated and the GEM stops.

If nonlinear, enter reasonable minimum and maximum values for your concentrations—These are the bounds from which the random samples are drawn for the random sampling iterative loop described above as part of the quasi-Newton algorithm. Only a single range is requested, which will be applied to the random sample drawn for *all* of the NEQN variables. Future versions of the GEM may include variable-specific bounds but only one is used for now. The range specified should be broad enough to include the expected final concentration values, yet narrow enough to lend real assistance to the iterative algorithm. (A starting guess that is orders of magnitude from the solution is not much assistance.) If expected final concentrations are near zero, we also advise the user to avoid specifying exactly zero as the lower bound, to avoid numerical problems. Instead, use a very small, but not precisely zero lower bound.

We conclude the description of the GEM input files by noting that there are numerous such files, some have somewhat complex structures, and opportunities for errors are numerous. Accordingly, Box 2.4 provides some guidance on minimizing errors when setting up these files for new problems.

**BOX 2.4   STEPS FOR SUCCESSFUL MODEL-
BUILDING OR ELEPHANT-EATING**

You've just received an interesting new assignment to build a new environmental model. After some brainstorming, you think it'll require 200 compartments, multiple state variables, some nonlinear kinetics and/or partitioning, and it's dynamic. Armed with the GEM and all good intentions, you populate the input files, develop the shell-required executables, and push the button. What happens?

If you're really lucky, you'll get an error message from the GEM that allows you to trace the problem. A little less lucky and you'll get a run-time overflow error that you have no idea how to solve. Even less lucky, the GEM will simply run into the ether and you'll have to reboot. What is almost certain, however, is that you will not have a successful run.

Why not? It's not you; it's the nature of model building. (If this were easy, your boss would do it.) Even if your theoretical construct is valid—and it's generally flawed at first—it seems inevitable that there will be data entry errors and/or some misunderstanding of the GEM's file formats. The "one small bite at a time" approach to eating elephants is the same strategy for model building. We provide the following suggestions.

If you ultimately want 200 compartments, start with 3 or 4. Ignore the kinetics. Ignore time-varying parameters/loadings and force your problem to be time-invariant. Start with one state variable. Ignore the sources and sinks. Check your input data files. Check them again. Have a colleague check them. Now, run that problem under the GEM's linear, steady-state mode and you stand a very good chance of a successful run. You may even be able to judge the validity of the result, e.g., the concentration may simply be the load divided by the flow. Now run that same time-invariant problem under a dynamic mode (suggest FTBS where you may be able to hand-verify the results at the first time step or two) at a small time step for many time steps, to run it out to steady state. Presumably, the two steady-state solutions will agree. Now start adding your complexity one small bite at a time.

If your problem is nonlinear, make it linear and follow the above steps. Then, run your linear problem under the GEM's nonlinear option. (Newton's method is equally valid for linear problems, it's just numerical overkill—but justified here for testing.) Use your known (linear) solution as the initial guess. Run at steady-state. Don't use a ridiculously small convergence criterion. Start with very small time steps. The GEM should converge immediately, because your guess is in fact the correct solution. Now change the guess slightly and run again. The GEM should converge in the second Newton's method iteration for a linear problem. Tighten up your convergence criterion. If you have nonlinear source/sink terms that will go in the $\underline{f(c)}$ executable, build your executable, but first use linear kinetics (first order) that you can check against with a linear run. (By having them in the VolumeSrcSnks.csv and AreaSrcSnks.csv.) Use a similar strategy for a nonlinear partitioning problem. If you are providing analytical derivatives to the Jacobian matrix and having problems, try using the finite difference numerical option (suggest 0.1 for derivative estimation parameter) or vice versa. Once you gain confidence that things are working, start adding your nonlinearities, again one small bite at a time.

Start preparing your client's invoice.

2.3   GEM OUTPUT FILES

GEM output files are described in this section. There are two types of output files: (1) those that contain the simulated concentrations and (2) those that contain intermediate diagnostic data. The simulated concentration files are named with the suffix .csv and are comma-separated value text files that may be read with any text editor or with a spreadsheet program if graphs are desired. The intermediate diagnostic files have the suffix .dng These are also text files and can be read with any text editor. We chose the hopefully obscure .dng suffix for these files *because the GEM erases all current *.dng files in the GEM folder immediately upon being executed.* That is done so that the .dng files to be written during runtime will always reflect the most current versions. If, for some reason, you have other files with this suffix in the same folder, they will be erased.

2.3.1   SIMULATED CONCENTRATION FILES

For *Environmental System* mode, simulated concentrations are reported in units of g/m^3. There are no internal unit conversion factors in the GEM. All inputs are in grams, meters, square meters, and cubic meters, resulting in g/m^3 output units.

For dynamic problems in *Environmental System* mode, there are three options for writing the concentration output files, as declared following the Control.csv record: *if dynamic and EqnS, enter option (1,2, or 3) for writing outputs.* Option 1 writes the output file ProfileSnapShot.csv that contains the concentrations of all state variables in all modeled compartments at a user-selected time step. The time step is entered following the Control.csv record: *if write option above = 1, enter the time step at which to print the profile.*

Option 2 writes the output file OneCompTimeSeries.csv that contains the concentrations of all state variables in a user-specified modeled compartment at a user-specified time step interval. The specified compartment and time step interval are entered following the Control.csv record: *if write option above = 2, enter the compartment number and the time step print interval.*

Option 3 writes the output file AllTimeSeries.csv. This file contains the concentrations of all state variables in all modeled compartments at all time steps. This is the most complete output format.

For steady-state problems in *Environmental System* mode, the concentrations are written by default to the ProfileSnapShot.csv file, as described above. For *Equation Solver* mode, the simulated results reflect the units of your inputs. For steady-state problems in *Equation Solver* mode, the results are written to the ProfileSnapShot.csv file. For dynamic results, the time series of results are written to the AllTimeSeries.csv file.

2.3.2   Intermediate Diagnostic Files

The diagnostic files provide information that is useful in model and data development. Several of these files essentially echo input data or results of internal calculations at a user-specified time step. The time step is entered following the Control.csv record: *if dynamic, enter the time step at which to output intermediate diagnostic files.* For steady-state problems, the desired time step defaults to zero. The files written at this user-specified time step are:

A.dng contains the elements of the **A** matrix.

b.dng contains the elements of the **b** vector.

RV.dng (for dynamic problems only) contains the main diagonal elements of the product of the **R** × **V** matrices. (The non-main diagonal elements are zero everywhere.)

FofC.dng contains the elements of the $\underline{f}(\underline{c})$ nonlinear function.

RofC.dng contains the main diagonal elements of the $\mathbf{R}(\underline{c})$ matrix. (The non-main diagonal elements are zero everywhere.)

The other diagnostic files are written to continuously during runtime as information becomes available. These are:

Error.dng contains error messages. The first error encountered during a run will trigger a run termination and, thus, the Error.dng file will contain only a message associated with that error. You may have multiple errors, but these will be revealed only sequentially in the Error.dng file over new runs, as previous errors are corrected.

EulerStabilityInfo.dng contains a listing of equations that violate the dynamic stability criterion and the times at which those violations occur. Very usefully, the maximum time step (days) that will just satisfy the stability criterion is also reported. This is applicable only to the FT (Euler) dynamic options.

Wiggle.dng contains a listing of adjacent compartments numbers for which the positivity/wiggle criterion is violated. This information is given by state variable and time. Very usefully, the maximum inter-compartment length that will just satisfy the positivity/wiggle criterion is also reported.

NewtonInfo.dng contains a listing of intermediate results during the quasi-Newton algorithm iterations. This is a fairly detailed list of results consisting of (at each time step): Newton iteration number, current x values, current f value, random sampling iteration number, lambda value, and the complete Jacobian matrix.

Warning: If you are making many time steps and/or encountering many Newton iterations, this file can become quite large. Because of this size issue, this file is written only if a Y is entered following the Control.csv record: *if nonlinear, would you like to save Newton Algorithm diagnostics to NewtonInfo.dng file? (Y,N)* Once your nonlinear problem works correctly for a few time steps, you should de-activate the writing of this file.

2.3.3 MASS BALANCE ERROR CHECKING

Although the GEM solves mass balance-based equations that by definition preserve mass, many opportunities for violations exist when developing and solving mass balance equations using approximate numerical methods for mass balance. One example is numerical dynamic instability. Another example is the use of large time steps in the presence of time-varying system parameters (see Box 2.5). Accordingly, we included an internal mass balance tracking algorithm that monitors the various

---

BOX 2.5   CHOICE OF TIME STEPS AND MASS BALANCE ERRORS

The underlying partial differential equations on which the GEM is based are mass balance equations; therefore, by definition, there is no intrinsic mass balance error. However, the way that the GEM is applied to any particular problem may result in mass balance errors. If you have essentially continuously varying loads and/or system parameters (e.g., flows, compartment volumes) and are making relatively large time steps, peaks and troughs in concentration may be missed with associated mass balance errors. Over the duration of a dynamic simulation, many of these errors will be self-correcting, but some error will occur. Chapter 6 has an extensive discussion of numerical errors that may result from time step choices (i.e., instability, numerical dispersion). In addition to these numerical considerations, the user is cautioned to use a time step that is commensurate with the time scale at which the problem has variability.

TABLE 2.21

MassBalance.csv

State Variable	Initial	In	Transport Out	Volume-Based Sinks Out	Area-Based Sinks Out	Residual	Balance Error
1	0	200000	138106.8	34975.46	0	26917.77	1.09E-10

fluxes of mass at each time step (the mass balance algorithm is implemented only for dynamic problems) that enter or leave the modeled compartments during a simulation and reports their cumulative values and the overall mass balance error among them at the conclusion of a simulation. Currently, the GEM monitors mass balance only for linear problems. The results are reported in the MassBalance.csv file. Specifically, the following mass fluxes are tracked:

1. The mass entering all internal (non-boundary or dummy) compartments exogenously due to a direct loading
2. The mass entering or leaving all internal compartments due to an endogenous volume-based source/sink (e.g., first-order decay)
3. The mass entering or leaving all internal compartments due to an endogenous area-based source/sink
4. The mass entering or leaving all internal compartments due to advective and dispersive transport from adjacent boundary or dummy compartments

For example, Table 2.21 is the MassBalance.csv output file resulting from the GEM simulation for the Chapter 1 PRAM-like model based on the F_d approach (Figure 1.6). Recall that that example involved a single state variable (dissolved PCB concentration), an external loading (PCB from the ship compartment), and an endogenous volume-based first-order decay process. The initial concentrations in all compartments were 0s. The cumulative loading (1,000 g/day × 200 days = 200,000 g) is the only input to the system (shown in the In column).

During the simulation, 138,106 g of PCB were transported out of the modeled system to boundary compartments (Transport Out column), while 34,975 g were lost internally due to decay. At the end of 200 days, the residual concentrations in the four internal compartments represented 26,917 g of PCB remaining in the system. Thus, the overall mass balance error for the simulation can be calculated as 200,000 − 138,106 − 34,975 − 26,917 or effectively 0. The last record reports the mass balance error as a fraction of total mass entering the system (200,000 g in this example) during the simulation.

REFERENCE

Mansfield, R. 2005. *Visual Basic 2005 Express Edition for Dummies*. Hoboken, NJ: Wiley Publishing.

3 Compartment Approach, Transport Mechanisms, and Boundary Conditions

3.1 COMPARTMENT APPROACH

Within the GEM's *Environmental System* mode, for dynamic problems the GEM builds and solves partial differential equations describing the concentrations of chemical constituents in environmental media in space and time subject to transport by advection and/or dispersion/diffusion, direct loadings, and internal sources and sinks. For example, for a single constituent in a single spatial dimension (x) and constant coefficients, the governing partial differential equation is

$$R\frac{\partial c}{\partial t} = v\frac{\partial c}{\partial x} + E\frac{\partial^2 c}{\partial x^2} \pm \text{sources}\Big/\text{sinks} \tag{3.1}$$

where
    c = concentration (M/L^3) dependent variable
    v = advective transport velocity coefficient (L/T)
    E = dispersive or diffusive transport coefficient (L^2/T)
    R = optional retardation term used in modeling porous media
    t = time (T) independent variable
    x = space (L) independent variable

A more general case may involve multiple interacting dependent variables and multiple spatial dimensions. The coefficients may be functions of space or time. The source and sink terms may also be functions of space or time and may be linear or nonlinear with respect to the dependent variables.

    Equation (3.1) is a second-order (level of highest derivative), non-homogeneous (containing source/sink terms), parabolic partial differential equation (Hoffman, 1992). Boundary and initial conditions must be considered and such problems are often termed propagation problems in that the solution is marched forward in time from an initial state and modified by boundary conditions at the edge of the domain of interest. This type of partial differential equation describes many environmental fate and transport problems of popular interest. In general, exact (analytical) closed-form solutions to these problems exist only for very simple situations, and numerical methods must be used instead. These numerical solutions are typically based on so-called finite difference approximations to the spatial and temporal derivatives.

This finite difference approach is used in the GEM; however, an important distinction is that the spatial derivatives are approximated not by finite differences to the equations themselves, but rather by discretizing the environmental media. These discretized media are conceptualized as compartments or volumes within which concentrations and environmental parameters (coefficients) are assumed spatially uniform. Under this compartment modeling approach (and other assumptions discussed later), Equation (3.1) can be approximated as

$$\frac{d(R_i V_i c_i)}{dt} = \sum_j Q_{i,j} c_{i,j} + \sum_j E'_{i,j}(c_j - c_i) \pm \text{sources}_i\Big/\text{sinks}_i \qquad (3.2)$$

where
    i = index for compartment i
    R_i = retardation coefficient for compartment i (unitless)
    V_i = volume of compartment i (L^3)
    $Q_{i,j}$ = advective flow rate across interface between compartments i and j (L^3/T)
    $E'_{i,j}$ = dispersive/diffusive flow rate across i,j boundary (L^3/T)
    c_i = chemical concentration in interior of compartment i (M/L^3)
    c_j = chemical concentration in interior of compartment j (M/L^3)
    $c_{i,j}$ = chemical concentration at interface between compartments i and j

The summations are over all compartments j that are contiguous neighbors of compartment i.

While Equation (3.2) appears to be an *ordinary* differential equation involving only a single independent variable (time), in fact the true underlying differential Equation (3.1) is a partial differential equation with time *and* space as independent variables. The compartment approach has in effect already discretized the partial differential equation in the space domain, thus moving us partially toward a full numerical solution. The solution to Equation (3.2) and analogous equations written for *all* compartments i can then be advanced in time using temporal finite difference methods.

For more than one compartment and/or more than one chemical constituent, the equation becomes a system of coupled differential equations that are also simultaneously advanced in time. At steady state (time derivative equals zero), we have a system of algebraic equations (linear or nonlinear) that can be solved using simultaneous equation numerical methods.

Transport is defined here as the movement of constituent mass between and among compartments by advective water movement (e.g., river flow) and/or dispersive-type fluxes (including macro-dispersion or molecular diffusion, as appropriate). Both mechanisms can be considered types of mass transport flow.

The GEM is not a hydraulic or hydrodynamic model. Advective flows and dispersion/diffusion coefficients between and among compartments at each time step (or steady-state condition) must be provided by the user. The remainder of this chapter describes how the transport terms are incorporated into the constituent mass balance equations, and outlines the matrix structure of the resulting system of simultaneous equations. The mass balance equations presented in this chapter include only the transport terms, external loadings, and boundary conditions, i.e., the constituent is environmentally conservative. This system of equations is augmented in Chapter 4 to include source and sink terms.

3.2 MASS BALANCE EQUATION FOR COMPARTMENT i

Consider an arbitrary compartment i that has zero, one, or more neighboring (contiguous) compartments j. Compartment i and its neighbors can have any spatial configuration (one, two or three dimensions). The compartment geometries are also essentially arbitrary so long as the interfacial areas (between adjoining compartments) and other relevant geometric parameters (described later) can be specified by the user.

The advective flow rate across an interface between compartments i and j is denoted as Q_{ij} with units of volume per time (L^3/T). If the constituent concentration at that interface is c_{ij}, then the advective mass transport from/to i to/from j is $Q_{ij}c_{ij}$ with units of mass/time (M/T). If we adopt the convention that flow is *positive* with respect to compartment i if the flow is *from* compartment j *to* compartment i (entering i) and *negative* if flow is *from* i *to* j (leaving i), a steady-state constituent mass balance around i due to advective transport only (assuming no internal sources or sinks of flow within i) is

$$\sum_j Q_{ij}c_{ij} = 0 \qquad (3.3)$$

Thus, the constituent mass advected into compartment i is balanced by the mass advected out at steady state.

Recall that our system of mass balance equations assumes that, with N compartments, there are N concentrations to be calculated, i.e., N equations in N unknowns. We need to express the concentration at the i,j interface (c_{ij}) as a function of concentrations in the *interiors* of i and j (c_i and c_j). Otherwise, we have more unknowns than equations. We do so by assuming that the concentration gradient between i and j is linear, so that the function is

$$c_{ij} = \alpha_{ij}c_i + (1 - \alpha_{ij})c_j \qquad (3.4)$$

where α_{ij} is a user-specified weighting factor (in the input file, Interfaces.csv) that can take on values between 0 and 1. For example, if the flow is from i into j and $\alpha_{ij} = 1$, the concentration at the interface is determined entirely by the upstream concentration c_i. For $\alpha_{ij} = 0$ and flow again from i to j, the concentration is determined entirely by the downstream concentration c_j. For $\alpha_{ij} = 0.5$, c_{ij} is effectively an average value between c_i and c_j. α_{ij} is analogous to a finite differencing parameter, i.e., a backward difference would use $\alpha_{ij} = 1$, a forward difference would use $\alpha_{ij} = 0$, and a central difference would use $\alpha_{ij} = 0.5$.

α_{ij} is user-specified for each i,j interface. If the user specifies a central difference (0.5) for i,j, then α_{ij} is calculated internally for that interface as

$$\alpha_{ij} = \frac{lc_{ji}}{lc_{ij} + lc_{ji}} \tag{3.5}$$

where

lc_{ij} = distance (L) from centroid of compartment i to i,j interface
lc_{ij} = distance (L) from centroid of compartment j to i,j interface

Thus, if compartments i and j are identically configured, $\alpha_{ij} = 0.5$. If not, α_{ij} becomes a weighted average.[*] (The weighting is such that the concentration is biased toward the closer centroid distance of the two compartments.)

We also need to carry forward our sign convention on flows (positive for compartment i if flow enters i and negative if flow leaves i. Therefore, we adopt the following[†]

$$c_{ij} = \begin{matrix} \alpha_{ij}c_i + (1-\alpha_{ij})c_j, \text{if } Q_{ij} < 0 \\ \alpha_{ij}c_j + (1-\alpha_{ij})c_i, \text{if } Q_{ij} > 0 \end{matrix} \tag{3.6}$$

Given a user-specified value of α_{ij}, we have expressed the advective transport between compartments in terms of concentrations in the contiguous compartment interiors.

Consider now mass transport between compartment i and neighboring compartment j due to a dispersive-type mechanism, where the driving force for mass transport is assumed to be the concentration gradient between the two compartments, i.e., $c_j - c_i$. (This is a common assumption and is based on Fick's law of diffusion.) If this gradient is positive, i.e., $c_j > c_i$, the dispersive transport is from j to i and vice versa.

Let E_{ij} be the dispersion (or diffusion) coefficient at the i,j interface with units of L^2/T. Let A_{ij} be the interfacial area (L^2), which is the area over which the dispersive flux acts. Let L_{ij} denote the distance (L) over which a concentration gradient is assumed to exist between compartments i and j. Then the variable

$$E'_{ij} = \frac{E_{ij}A_{ij}}{L_{ij}} \tag{3.7}$$

has units of (L^3/T) and can be thought of as a flow due to dispersive (or diffusive) transport. With this dispersive flow rate and the concentration gradient, the mass transport between compartments i and j due to dispersion is $E'_{ij}(c_j - c_i)$.

Analogous to our advective transport convention, if the gradient is into i ($c_j > c_i$), the above mass transport term has a positive value, and vice versa. Therefore, the

---

[*] If a user specifically wants to use a constant 0.5 for every α_{ij} regardless of relative compartment configurations and sizes, the GEM's weighting method can be circumvented by entering a value that, for all practical purposes, is 0.5 but is not *exactly* 0.5, e.g., 0.499.

[†] To illustrate the need for the flow direction convention, if we wanted a backward differencing scheme ($\alpha_{ij} = 1$), Equation (3.6) becomes

$$c_{ij} = \begin{matrix} c_i, \text{if } Q_{ij} < 0 \\ c_j, \text{if } Q_{ij} > 0 \end{matrix}$$

which is exactly what we want.

(steady-state) mass balance around compartment i due only to dispersion from the neighboring compartments is

$$\sum_j E'_{ij}(c_j - c_i) = 0 \qquad (3.8)$$

To facilitate writing all these mechanisms into a single mass balance equation, given the somewhat awkward flow convention, we now introduce a binary $(0 - 1)$, integer variable I_{ij} such that

$$I_{ij} = \begin{array}{l} 1, \text{ if } Q_{ij} < 0 \\ 0, \text{ if } Q_{ij} > 0 \end{array} \qquad (3.9)$$

With this binary variable, we can now combine these mechanisms into the general time-varying differential equation that represents constituent mass balance for compartment i due to advective and dispersive transport mechanisms. Combining the above equations, assuming time-varying conditions, and including a variable representing an external loading of constituent into compartment $i - W_i$ (M/T), we have the following mass balance equation for compartment i

$$d\frac{(V_i c_i)}{dt} = \sum_j Q_{ij}[I_{ij}(\alpha_{ij}c_i + (1 - \alpha_{ij})c_j) + (1 - I_{ij})(\alpha_{ij}c_j + (1 - \alpha_{ij})c_i)]$$

$$+ \sum_j E'_{ij}(c_j - c_i) + W_i \qquad (3.10)$$

Equation (3.10) can be expanded as

$$d\frac{(V_i c_i)}{dt} = c_i \sum_j \alpha_{ij}Q_{ij}I_{ij} + \sum_j (1 - \alpha_{ij})Q_{ij}I_{ij}c_j + \sum_j \alpha_{ij}Q_{ij}(1 - I_{ij})c_j$$

$$+ c_i \sum_j (1 - \alpha_{ij})Q_{ij}(1 - I_{ij}) - c_i \sum_j E'_{ij} + \sum_j E'_{ij}c_j + W_i \qquad (3.11)$$

and grouping coefficients of c_i and the various c_j values can be rearranged as

$$d\frac{(V_i c_i)}{dt} = c_i \sum_j [\alpha_{ij}Q_{ij}I_{ij} + (1 - \alpha_{ij})Q_{ij}(1 - I_{ij}) - E'_{ij}]$$

$$+ \sum_j c_j[(1 - \alpha_{ij})Q_{ij}I_{ij} + \alpha_{ij}Q_{ij}(1 - I_{ij}) + E'_{ij}] + W_i \qquad (3.12)$$

A simple example illustrates Equation (3.12) for a single compartment in Box 3.1.

BOX 3.1   ILLUSTRATION OF TRANSPORT TERMS

Let compartment i be bordered by four adjacent compartments numbered 1 through 4 as shown in Figure 3.1. Flow is from compartment 1 into compartment i, and from compartment i into each of compartments 2 through 4. First, we use a backward difference, i.e., $\alpha_{ij} = 1$ for all four interfaces.

Recall that flows entering compartment i are considered positive and those leaving are negative (and entered by the user as negative). Then, from Equation (3.9), $I_{i1} = 0$ while $I_{i2} = I_{i3} = I_{i4} = 1$. Equation (3.12) then becomes for the backward difference

$$d\frac{(V_i c_i)}{dt} = c_i[-E'_{i1} - Q_{i2} - E'_{i2} - Q_{i3} - E'_{i3} - Q_{i4} - E'_{i4}]$$

$$+ c_1(E'_{i1} + Q_{i1}) + c_2 E'_{i2} + c_3 E'_{i3} + c_4 E'_{i4} + W_i$$

where we are now showing the negativity of the flows leaving compartment i explicitly for clarity. The terms on the right side that are coefficients of c_i are advective transport and dispersive *out* of compartment i into the other four compartments. The terms that are coefficients of c_1, c_2, c_3, or c_4 are advective and dispersive transport into compartment i.

It is more illuminating to rearrange this equation to show how each of the interface flow terms constitutes sources or sinks to compartment i by writing it as

$$d\frac{(V_i c_i)}{dt} = Q_{i1}c_1 - Q_{i2}c_i - Q_{i3}c_i - Q_{i4}c_1 + E'_{i1}(c_1 - c_i)$$

$$+ E'_{i2}(c_2 - c_i) + E'_{i3}(c_3 - c_i) + E'_{i4}(c_4 - c_i) + W_i$$

where we see that the advective transfer terms involve only the upstream concentration because of the backward difference. For the central differencing option, $\alpha_{ij} = 0.5$, the comparable equation becomes

$$d\frac{(V_i c_i)}{dt} = 0.5Q_{i1}(c_1 + c_i) - 0.5Q_{i2}(c_2 + c_i) - 0.5Q_{i3}(c_3 + c_i) - 0.5Q_{i4}(c_4 + c_1)$$

$$+ E'_{i1}(c_1 - c_i) + E'_{i2}(c_2 - c_i) + E'_{i3}(c_3 - c_i) + E'_{i4}(c_4 - c_i) + W_i$$

where we now see that the advective transport terms involve an average of the concentrations on *both* sides of the interface.

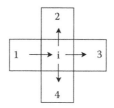

FIGURE 3.1 Compartment orientation.

3.3 SYSTEM OF TRANSPORT MASS BALANCE EQUATIONS FOR SINGLE CHEMICAL

This section is where we need to begin using matrix notation to describe the GEM set of mass balance equations. It is impractical to try to describe these simultaneous equations otherwise. We assume that the reader has some basic understanding of matrix notation; nonetheless, we have included an overview in the Appendix.

Consider now the system of differential equations that results from applying the mass balance Equation (3.12) to *all* compartments *i* that comprise the spatial domain modeled, but first for a single state variable (chemical) only. The larger system of simultaneous equations arising from multiple chemicals is described in the following section.

Let NIC be the number of interior[*] compartments. Writing Equation (3.12) for all NIC interior compartments, $i = 1, ..., $ NIC, we can express the resulting system of equations as the matrix equation

$$\frac{d(\mathbf{V}\underline{c})}{dt} = \mathbf{A}\underline{c} + \underline{b} \tag{3.13}$$

where

$\underline{c}$ = NIC × 1 vector of constituent concentrations, i.e., c_i, $i = 1, ..., $ NIC. (We use **bold** notation to indicate matrices and vectors in this document. Vectors are further identified by lower case font and underscoring.)

$\mathbf{V}$ = is the NIC × NIC diagonal matrix with the compartment volumes V_i, $i = 1, ..., $ NIC on the main diagonal and 0s elsewhere.

$\mathbf{A}$ = NIC × NIC matrix of coefficients of the concentrations. The main diagonal elements of $\mathbf{A}$ are given by the terms that are coefficients of c_i in the right side of Equation (3.12). The off-diagonal elements are given by the coefficients of the c_j terms in the right side of Equation (3.12), appropriately modified for boundary conditions, to be discussed in Section 3.6.

$\underline{b}$ = NIC × 1 vector of external inputs and boundary conditions, i.e., $b_i = W_i +$ the coefficients of the c_j terms in the right side of Equation (3.12) that are fixed boundary compartments, as discussed in Section 3.6.

---

[*] An interior compartment is not a boundary compartment or dummy compartment (explained later). Predictions are made for chemicals within interior compartments.

3.4 SYSTEM OF TRANSPORT MASS BALANCE EQUATIONS FOR MULTIPLE CHEMICALS

Consider now having NSV state variables (chemicals) subject again to transport mechanisms only. We need to update the mass balance equation for compartment i previously expressed for a single chemical (Equation 3.12). We denote the concentration of chemical k in compartment i as $c_{i,k}$. The mass balance equation for chemical k in compartment i due to transport only is then

$$d\frac{(V_i c_{i,k})}{dt} = c_{i,k} \sum_j [\alpha_{ij} Q_{ij} I_{ij} + (1 - \alpha_{ij}) Q_{ij} (1 - I_{ij}) - E'_{ij}]$$

$$+ \sum_j c_{j,k} [(1 - \alpha_{ij}) Q_{ij} I_{ij} + \alpha_{ij} Q_{ij} (1 - I_{ij}) + E'_{ij}] + W_{i,k}$$

(3.14)

where it should be noted that the external loading term ($W_{i,k}$) has also been updated with the k index to reflect that it is a loading specific to chemical k.[*]

For NSV chemicals, there are NSV equations, i.e., $k = 1, \ldots,$ NSV, analogous to Equation (3.13) for each interior compartment i. Thus, the system of equations potentially increases from NIC equations (one chemical) to NIC*NSV equations.[†]

As we saw in the Chapter 1 PRAM-like introductory example, however, it is possible to have different state variables relevant to different compartments. We denote the maximum number of state variables being simultaneously modeled in any single compartment as MaxNSV. We will also denote the number of equations that pertain to an individual state variable k as NEQN_k. NEQN_k is the number of interior compartments relevant to state variable k. Let NEQN denote the total number of equations across all state variables and all interior compartments. If all MaxNSV state variables pertain to all NIC compartments, then NEQN = MaxNSV*NIC. In the more general case, however, with compartment-varying state variables,

$$NEQN = \sum_{k=1}^{MaxNSV} NEQN_k$$

The matrix structure of the larger simultaneous system of NEQN equations analogous to the matrix equation (3.13) is designated as

$$\frac{d(\mathbf{V'\underline{c}'})}{dt} = \mathbf{A'\underline{c}' + \underline{b}'}$$

(3.15)

---

[*] In the most general case, there could be chemical-specific flow and/or dispersive/diffusive coefficients and Q_{ij} and E'_{ij} should also have a k index. Indeed, the GEM allows for this generality in the Flows.csv and ECoefficients.csv input files. To maintain relative simplicity in this document, however, we are assuming that transport parameters apply equally to all k chemicals.

[†] Because sources and sinks among the chemicals are not included here (see later chapters), the system of transport equations across multiple chemicals could be solved independently for each chemical, i.e., there is no need to couple them and solve them simultaneously. However, we consider them simultaneously here as a prelude to the later treatment in which they are coupled via source/sink terms.

$$\begin{bmatrix} [A_{1,1}'] & & & \\ & [A_{2,2}'] & & \\ & & \text{and so on ...} & \\ & & & [A_{MaxNSV,MaxNSV}'] \end{bmatrix} \begin{bmatrix} [\underline{c}_1'] \\ [\underline{c}_2'] \\ \text{and so on...} \\ [\underline{c}_{MaxNSV}'] \end{bmatrix} + \begin{bmatrix} [\underline{b}_1'] \\ [\underline{b}_2'] \\ \text{and so on...} \\ [\underline{b}_{MaxNSV}'] \end{bmatrix}$$

FIGURE 3.2   Matrix structure of system of mass balance equations (transport only).

where the prime indicates a larger matrix or vector than the **V**, **A**, $\underline{c}$, and $\underline{b}$ matrices and vectors used previously in Equation (3.13). (It does *not* indicate the matrix transpose operation as is often used in matrix algebra notation.) Thus, matrices **V′** and **A′** have dimension NEQN × NEQN while vectors $\underline{c}'$ and $\underline{b}'$ have dimension NEQN × 1.

For our transport-only assumption, the structure of the new matrices and vectors is a straightforward extension of the single chemical structure. The $\underline{c}'$ and $\underline{b}'$ vectors consist of the NEQN$_1$ elements of vectors $\underline{c}$, and $\underline{b}$, respectively, from matrix Equation (3.13) for the first chemical as their first elements, the analogous NEQN$_2$ elements of vectors $\underline{c}$, and $\underline{b}$ for the second chemical as their next elements, and so on sequentially through the MaxNSV chemicals. Analogously, matrices **V′** and **A′** consist of MaxNSV blocks made of the chemical-specific **V** and **A** matrices from Equation (3.13). The **A′**, $\underline{c}'$, and $\underline{b}'$ matrices and vectors are illustrated in Figure 3.2. The subscript notation refers to the chemicals. That is, $A'_{k,k}$ is the NEQN$_k$ × NEQN$_k$ sub-matrix of matrix **A′** pertaining to chemical k. Similarly, $\underline{c}'_k$ and $\underline{b}'_k$ refer to NEQN$_k$ × 1 sub-vectors of vectors $\underline{c}'$ and $\underline{b}'$, respectively, pertaining to chemical k.

The sub-matrices on the main diagonal of **A′** are referred to as main diagonal sub-matrices. Sub-matrices off the main diagonal of **A′** are referred to as off-main diagonal sub-matrices. All off-main-diagonal sub-matrices elements are 0s because there is no between-chemical coupling under the transport-only scenario as discussed above. (This will change when we add source/sink terms in Chapter 4.)

Matrix **V′** has an even simpler structure that will remain true even as we add source/sink terms later. Its non-zero elements are only on the main diagonal (i.e., it is a diagonal matrix) and consist of simply repeating the compartment volumes for each chemical as one walks down the main diagonal. Thus, using an analogous notation for the **A′** sub-matrices, the kth sub-matrix of **V′**, i.e., $V'_{k,k}$ corresponding to arbitrary chemical k, is identically equal to the original NEQN$_k$ × NEQN$_k$ **V** matrix defined as part of the matrix Equation (3.13).

In the remainder of this document, we drop the prime (′) superscript from the **V′**, **A′**, $\underline{c}'$, and $\underline{b}'$ matrices/vectors for simplicity. Hereafter, we will agree that **V**, **A**, $\underline{c}$, and $\underline{b}$ refer to the most general scenario of fate and transport involving possibly multiple chemicals.

3.5   EXTENSION TO POROUS MEDIA

The development of Equation (3.15) implicitly assumed that the chemical concentration in compartment i, c_i reflects the entire compartment volume V_i. For porous media (groundwater systems), we want to define c_i to be the *solute* concentration, i.e., chemical mass concentration per unit volume of *pore water*, not the concentration per unit volume of water and solids.

To achieve this we must identify the GEM parameters affected by this change and make the appropriate adjustment. Those parameters are obviously volume and area. We adopt the convention that the flow rate per se is not affected by the presence of solids; the user-supplied flow rate is assumed to be valid for the particular system modeled. Rather, the flow velocity through the pore spaces would obviously be greater (for the same flow rate) in porous media than in pure aqueous systems. The flow velocity, however, is not an explicit parameter of the GEM.

Therefore, we need to multiply the volume parameter in the derivative term and the area parameter in the dispersion term (and, later, the volume-based and area-based source/sink terms) by the water content Θ, which has units of L^3 (water)/L^3 (total). The equation for mass transport in compartment i for is now[*]

$$d \frac{(V_i \theta_i c_{i,k})}{dt} = c_{i,k} \sum_j [\alpha_{ij} Q_{ij} I_{ij} + (1 - \alpha_{ij}) Q_{ij} (1 - I_{ij}) - \theta_i E'_{ij}]$$

$$+ \sum_j c_{j,k} [(1 - \alpha_{ij}) Q_{ij} I_{ij} + \alpha_{ij} Q_{ij} (1 - I_{ij}) + \theta_i E'_{ij}] + W_{i,k}$$

(3.16)

and the generalized matrix equation can be expressed as

$$\frac{d(\mathbf{V\theta c})}{dt} = \mathbf{A}(\theta)\underline{c} + \underline{b}$$

(3.17)

where we continue to drop the prime symbols ($'$) as discussed above and Θ is a NEQN × NEQN main diagonal matrix with MaxNSV-repeating blocks of Θ_i, $i = 1$, ..., NEQN$_i$ for its main diagonal elements and 0s elsewhere, similar to the structure described above for matrix $\mathbf{V}$. It is also understood that matrix A now contains the water content parameters that are coefficients of the dispersion terms. The forcing function $\underline{b}$ remains as before, because the external loadings are simply as described— external loadings—and are not affected by the volume to which they apply. Thus, for example, if the modeled system were a porous medium and c_i is solute chemical concentration with units of M/L^3_{water}, V_i is total compartment volume (L^3_{total}), and A_i is total cross-sectional area (L^2_{total}), then the units of the derivative and dispersive transport terms are consistent (M/T) given the L^3_{water}/L^3_{total} units (for volume) or L^2_{water}/L^2_{total} (for area) of the water content parameter.

Matrix Equation (3.17) is now the most generalized form of the system of equations as we go forward and it will be understood that the user will define Θ as

---

[*] Notice that we could simplify this equation somewhat by dividing through by Θ so that the only terms in which Θ appears are the advective flow terms where Q is divided by Θ. In Chapter 4, in the context of a retardation sink we will do this. For now, we leave the equation as above.

appropriate to the system being modeled. For example, for surface water modeling where it is assumed that the water content is effectively 1.0, Θ would be the NEQN × NEQN identify matrix, $\mathbf{I}$ (values of 1 on the main diagonal and 0s elsewhere) reflecting a purely aqueous system.

3.6   BOUNDARY CONDITIONS AND TYPES OF COMPARTMENTS

Before discussing boundary conditions, a brief discussion of the types of compartments recognized by the GEM is needed. The three types of compartments in the GEM are interior compartments, boundary compartments, and dummy compartments. An interior compartment denotes a compartment in which concentrations are to be simulated, i.e., modeled. Each state variable in each interior compartment represents one mass balance equation to be solved. If a compartment is not an interior compartment, it is either a boundary compartment or a dummy compartment (both sometimes called exterior compartments). Boundary conditions in boundary compartments determine how the concentrations in the set of interior compartments modeled by the GEM are affected by the outside world. Influences from outside on modeled results are included by boundary conditions that apply within GEM boundary compartments. Boundary compartment conditions affect conditions within the modeled, interior compartments and are discussed in detail later in this section.

Dummy compartments are, for GEM purposes, neither interior modeled compartments nor boundary condition compartments. Rather, they are bookkeeping conveniences that allow cross-compartment flows or mass fluxes leaving a modeled interior compartment to the outside world, *when the outside world does not affect the interior compartment*, a place "to go to." Dummy compartments are not necessitated by the mathematics of the problem but rather to accommodate the GEM architecture and input file formats.

For example, consider the most downstream compartment in a system involving plug flow (no dispersion) and where a backward difference ($\alpha = 1$) is used. For example, in Figure 3.1, consider compartment 3. The flow leaving interior compartment i in Figure 3.1 and shown as entering compartment 3 needs a "go-to" location to satisfy the GEM architecture. That place would be provided by a dummy compartment 3. Note that in this scenario, concentrations in compartment 3 do not affect concentrations in compartment i in any way.

No physical dispersion across the $i,3$ interface is assumed and, in addition, the backward difference prevents any numerical averaging across this interface. Nonetheless, we need to provide dummy compartment 3 to accommodate the GEM design. Another example is where an area-based sink (e.g., volatilization or settling; see Chapter 4) results in a mass flux transfer from a modeled interior compartment to the outside world (that again does not affect the modeled interior compartment). Suppose compartment i in Figure 3.1 involved a volatilization loss across the $i,3$ interface. In this case, the GEM needs access to the $i,3$ interface area to compute the transfer flux. Therefore, compartment 3 must exist as a dummy compartment to have this interfacial area parameter available.

The number* of interior (modeled) compartments is again NIC. The number of exterior boundary compartments is NBC and the number of dummy exterior compartments is NDC. Thus, the total number of compartments is NT = NIC + NBC + NDC.

Moving on to the important topic of boundary conditions, it is not an exaggeration to say that the proper use of boundary conditions is imperative in developing a credible model. Too often, boundary conditions are assumed that simply are not appropriate and the resulting model is invalid. Indeed, the interplay and tradeoffs between (1) knowing which boundary conditions are appropriate and when and where they should be used and (2) the appropriate spatial domain of the model (vis-á-vis the outside world) constitute a very large part of the art of model development.

Three general categories of boundary conditions for partial differential equations are of interest for the GEM: Dirichlet-type boundary conditions, wherein boundary concentrations are specified (fixed), Neumann-type boundary conditions where concentration derivatives (gradients) are specified, and mixtures of the two. Among these categories, the GEM includes three types of boundary conditions: (1) fixed concentration (Dirichlet), (2) zero gradient (Neumann), and (3) linear gradient (Neumann). The implementation of these types of boundary conditions into the system of mass balance equations and suggestions on when they are appropriate to use are described below.

3.6.1 Fixed Concentration Boundary Conditions

This boundary condition specifies a fixed concentration in the boundary compartment. (It is fixed at a given time step for dynamic problems. It can vary temporally, however.) Such a boundary compartment is appropriate when it can be reasonably assumed that the simulated concentrations for the GEM's interior compartments do not affect concentrations in the boundary compartments. A prime example of this is when the boundary compartment is the most upstream compartment in a strongly advective flow system. A downstream example would be when the system modeled is discharging into a very much larger volume within which those boundary concentrations can reasonably be assumed to not be affected by the discharge, e.g., a large completely mixed lake or the ocean. We note that use of a fixed concentration boundary condition imposes a very stringent assumption on a model. Make sure it is appropriate.

Equation (3.16) is our mass balance equation for interior compartment i corresponding to mass transport and an external loading. (Source and sink terms covered in Chapter 4 do not affect boundary conditions.) The transport terms in this equation are either coefficients of the concentration in compartment i, i.e., c_i, or coefficients of the concentrations in neighboring compartments c_j. Suppose that one of these neighboring compartments, say compartment 1, is a boundary compartment with a fixed concentration c_l. We can then write Equation (3.16) for a single chemical (boundary conditions are chemical-specific) explicitly distinguishing the transport terms that are coefficients of the boundary condition c_l as

---

* Compartments must be consecutively numbered with integers beginning with 1. Other than that restriction, interior and exterior compartments can be arbitrarily numbered.

$$d\frac{(V_i\theta_i c_i)}{dt} = c_i \sum_j [\alpha_{ij}Q_{ij}I_{ij} + (1-\alpha_{ij})Q_{ij}(1-I_{ij}) - \theta_i E'_{ij}]$$

$$+ \sum_{j \neq l} c_j[(1-\alpha_{ij})Q_{ij}I_{ij} + \alpha_{ij}Q_{ij}(1-I_{ij}) + \theta_i E'_{ij}] \qquad (3.18)$$

$$+ \{W_i + c_l[(1-\alpha_{il})Q_{il}I_{il} + \alpha_{il}Q_{il}(1-I_{il}) + \theta_i E'_{il}]\}$$

Recall that the loading W_i is a known value. In addition, the fixed boundary condition, c_l, and all of the transport factors affecting c_l are also known values. Thus, the implementation of the fixed boundary condition in the GEM simply considers the forcing function for compartment i to be the last two terms in Equation (3.18), i.e., the loading and the mass transport terms affecting the boundary condition. Equation (3.18) is applicable when compartment i is adjacent to a fixed concentration-type boundary compartment l.

3.6.2 ZERO GRADIENT BOUNDARY CONDITIONS

If one is not comfortable making the fixed concentration boundary assumption, a somewhat less restrictive assumption is available by using a zero gradient condition. The underlying assumption is that you don't know the concentration in the boundary compartment; otherwise you'd use a fixed concentration boundary. But you are comfortable in believing that the concentration in that boundary is essentially no different from the modeled concentration in the adjacent interior compartment. Typically, such an assumption would be made when the effects that your model is being developed to simulate or predict are relatively localized within your model and sufficiently far from those interior (modeled) compartments that abut boundary compartments as to have little effect on the outside world.

Consider Equation (3.18) above where the index l denotes the boundary compartment. The transport terms that are coefficients of c_l are given in the last term on the right side. For the zero gradient boundary condition, by definition, c_i will equal c_l when compartment i is adjacent to boundary compartment l. Because $c_i = c_l$, we can write Equation (3.18) as

$$d\frac{(V_i\theta_i c_i)}{dt} = c_i \left\{ \sum_j [\alpha_{ij}Q_{ij}I_{ij} + (1-\alpha_{ij})Q_{ij}(1-I_{ij}) - \theta_i E'_{ij}] \right.$$

$$\left. + [(1-\alpha_{il})Q_{il}I_{il} + \alpha_{il}Q_{il}(1-I_{il}) + \theta_i E'_{il}] \right\} \qquad (3.19)$$

$$+ \sum_{j \neq l} c_j[(1-\alpha_{ij})Q_{ij}I_{ij} + \alpha_{ij}Q_{ij}(1-I_{ij}) + \theta_i E'_{ij}] + W_i$$

Expanding the terms on the right side that are coefficients of c_i we can write

$$
\begin{aligned}
d\frac{(V_i\theta_i c_i)}{dt} = c_i \Bigg\{ &\sum_{j\neq l}[\alpha_{ij}Q_{ij}I_{ij}+(1-\alpha_{ij})Q_{ij}(1-I_{ij})-\theta_i E'_{ij}] \\
&+[\alpha_{il}Q_{il}I_{il}+(1-\alpha_{il})Q_{il}(1-I_{il})-\theta_i E'_{il}] \\
&+[(1-\alpha_{il})Q_{il}I_{il}+\alpha_{il}Q_{il}(1-I_{il})+\theta_i E'_{il}]\Bigg\} \\
&+\sum_{j\neq l}c_j[(1-\alpha_{ij})Q_{ij}I_{ij}+\alpha_{ij}Q_{ij}(1-I_{ij})+\theta_i E'_{ij}]+W_i
\end{aligned}
\tag{3.20}
$$

Of the several transport terms on the right side involving index l, all of these cancel except for Q_{il}. Thus, this equation greatly simplifies to

$$
\begin{aligned}
d\frac{(V_i\theta_i c_i)}{dt} = c_i \Bigg\{ &\sum_{j\neq l}[\alpha_{ij}Q_{ij}I_{ij}+(1-\alpha_{ij})Q_{ij}(1-I_{ij})-\theta_i E'_{ij}]+[Q_{il}]\Bigg\} \\
&+\sum_{j\neq l}c_j[(1-\alpha_{ij})Q_{ij}I_{ij}+\alpha_{ij}Q_{ij}(1-I_{ij})+\theta_i E'_{ij}]+W_i
\end{aligned}
\tag{3.21}
$$

Equation (3.21) is applicable when compartment i is adjacent to a zero concentration gradient-type boundary compartment l. This equation makes perfect sense. For a zero gradient boundary condition in compartment l, the only transport mechanism is advective flow. There is no concentration gradient; therefore, no dispersive transport. When Q_{il} is negative (flow from i into l), the result is a loss of mass from i. When Q_{il} is positive (flow from l into i), it is a source of mass from l.

3.6.3 LINEAR GRADIENT BOUNDARY CONDITIONS

Recall from above that the underlying assumption of the zero gradient boundary condition is that you don't know what the concentration in the boundary compartment is, but you are comfortable in believing that the concentration in that boundary is essentially no different than the modeled concentration in the adjacent interior compartment. This assumption can be relaxed even further by the linear gradient boundary condition, which is a very interesting mathematical device for essentially throwing your hands up and admitting that you don't know what the boundary concentration is (fixed) and are not sure that the concentrations in the extremities of your modeled domain can be reasonably assumed to be very similar to the outside world (zero gradient).

The linear gradient boundary condition makes the assumption that the concentrations in the boundary compartment 1 and those in the nearby interior compartments are not equal but are linearly related. Under this assumption, a linear extrapolation is made from the modeled nearby compartments to infer what the concentration in the boundary compartment is, and then that inferred concentration is essentially treated as a fixed boundary condition. Such an assumption would be most appropriate when the concentrations within the outlying interior compartments can be assumed to not differ greatly and/or when the outlying compartments are relatively small in size so that, even if there are strong gradients, they are approximately linear.

The linear gradient approach is easiest to understand by considering a one-dimensional system, for example, a river with concentrations varying only along the downstream direction. Let us consider an arbitrary compartment i in such a system, with adjacent upstream compartment $j1$ and adjacent downstream compartment $j2$. (We are trying to maintain the indexing scheme used previously where j is an index over all compartments that are contiguous to the current compartment under consideration i. Thus, j consists of $j1$ and $j2$.) A plot of concentration over the river sections consisting of $j1$, i, and $j2$ might look like Figure 3.3. The $j1$, i, and $j2$ points on the horizontal axis (distance downstream) indicates the locations of the centroids of those compartments.

Assuming a linear relationship among the three concentrations, from similar triangles we can write

$$\frac{c_{j1} - c_i}{L_{i,j1}} = \frac{c_i - c_{j2}}{L_{i,j2}} \qquad (3.22)$$

Assume that compartment $j2$ is a boundary compartment for which we want a linear gradient boundary condition. Solving Equation (3.22) for our unknown boundary compartment concentration, we get

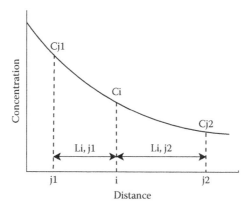

FIGURE 3.3   One-dimensional system.

$$c_{j2} = -\frac{L_{i,j2}}{L_{i,j1}} c_{j1} + \left(1 + \frac{L_{i,j2}}{L_{i,j1}}\right) c_i \qquad (3.23)$$

We can now treat the estimated value of c_{j2} as a fixed concentration boundary condition. The implementation of this, however, is most easily performed implicitly by suitably modifying the transport coefficients affecting the interior compartments $j2$ and i. For our one-dimensional system, the governing Equation (3.16) for a single chemical is

$$\frac{d(V_i\theta_i c_i)}{dt} = c_i a_{i,i} + c_{j1} a_{i,j1} + c_{j2} a_{i,j2} + W_i \qquad (3.24)$$

where $a_{i,i}$, $a_{i,j1}$, and $a_{i,j2}$ are the elements of the **A** matrix of transport coefficients in the row corresponding to concentration c_i. For example, $a_{i,j1}$ contains the transport terms affecting transport between i and $j1$. Combining Equations (3.23) and (3.24), results in

$$\frac{d(V_i\theta_i c_i)}{dt} = c_i \left(a_{i,i} + a_{i,j2} \left(1 + \frac{L_{i,j2}}{L_{i,j1}}\right) \right) + c_{j1} \left(a_{i,j1} - a_{i,j2} \frac{L_{i,j2}}{L_{i,j1}} \right) + W_i \qquad (3.25)$$

and the corresponding elements of the **A** matrix would be modified as shown above to implement the linear gradient boundary condition. If we had assumed that the upstream compartment $j1$ were the boundary compartment instead of $j2$, an analogous equation would result. Indeed, we can express Equation (3.25) more generally as

$$\frac{d(V_i\theta_i c_i)}{dt} = c_i \left(a_{i,i} + a_{i,B} \left(1 + \frac{L_{i,B}}{L_{i,ANB}}\right) \right) + c_{ANB} \left(a_{i,ANB} - a_{i,B} \frac{L_{i,B}}{L_{i,ANB}} \right) + W_i \qquad (3.26)$$

where index B denotes the boundary compartment and ANB denotes the adjacent non-boundary compartment.

Let us return now to the more general topological case where the ith compartment of interest has than one *or more* contiguous, non-boundary or non-dummy (interior) compartments. We assume that there remains only a single boundary compartment adjacent to compartment i.* For the single adjacent interior compartment scenario it was straightforward to develop the linear gradient boundary condition. For multiple

---

* While the GEM may be configured so that compartment i is contiguous to multiple boundary compartments, if one is using linear gradient boundary conditions (the use of which acknowledges uncertainty of the true concentrations in those boundary compartments as discussed above), it would seem pointless to do so.

adjacent interior compartments, while there are several alternative ways to incorporate this complexity (e.g., find the closest adjacent compartment in terms of distance to centroid and use it), we have adopted an approach that uses the *average* concentrations in the adjacent interior compartments and the *average* lengths between the adjacent interior compartments and compartment i. With this modification, Equation (3.23) can be expressed as

$$c_B = -\frac{L_{i,B}}{\overline{L}_{i,ANB}} \overline{c}_{ANB} + \left(1 + \frac{L_{i,B}}{\overline{L}_{i,ANB}}\right)c_i \tag{3.27}$$

where we are again using B as the boundary compartment index and ANB as the adjacent non-boundary index. $\overline{L}_{i,ANB}$ is the average length from the centroid of compartment i to all adjacent non-boundary compartments. Similarly, $\overline{c}_{ANB}$ is the average concentration of the adjacent, non-boundary compartments. Substituting Equation (3.27) into our overall governing equation (3.16), for a single chemical, we get, analogous to the one-dimensional version above (Equation 3.26), the following result

$$\frac{d(V_i\theta_i c_i)}{dt} = c_i\left(a_{i,i} + a_{i,B}\left(1 + \frac{L_{i,B}}{\overline{L}_{i,ANB}}\right)\right) + \sum_{j=1}^{NANB} c_j\left(a_{i,j} - \frac{a_{i,B}}{NANB}\frac{L_{i,B}}{\overline{L}_{i,ANB}}\right) + W_i \tag{3.28}$$

where $NANB$ is the number of non-boundary compartments adjacent to compartment i.

EXAMPLE 3.1

Comparison of Boundary Conditions

We illustrate the fixed, zero gradient, and linear gradient boundary conditions with an example using a one-dimensional river with advection, dispersion, and a first-order linear sink. (Sinks are discussed in Chapter 4.)

This example assumes a constant cross-sectional area of the river of 200 m² and a constant flow of 1,000 m³/day. We compartmentalized the river into 21 compartments, each 100 m long. Compartments 1 (most upstream) and 21 (most downstream) are boundary compartments. Compartment 1 has a fixed concentration boundary condition of 10 g/m³. We specified alternately, zero gradient, linear gradient, and fixed concentration boundary conditions for compartment 21. The fixed concentration lower boundary condition was arbitrarily specified at 3 g/m³. A longitudinal dispersion coefficient E of 1,000 m²/day was assumed as well as a first-order loss (decay) coefficient of $k = 0.01$/day. The water content is 1.0.

We ran the GEM under steady-state mode using a central difference ($\alpha_{ij} = 0.5$) for all interfaces for each of the three lower-boundary condition scenarios.

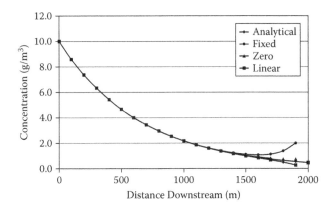

FIGURE 3.4a   Comparison of boundary conditions.

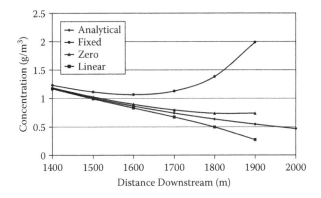

FIGURE 3.4b   Close-up view of Figure 3.4a.

In addition, we calculated the exact concentration profile using the analytical solution:

$$c(x) = c_0 e^{\frac{vx}{2E}(1-a)}$$

where

c_0 = upstream (boundary) concentration

v = velocity

$$a = \sqrt{1 + \frac{4kE}{v^2}}$$

k = first-order sink

The three GEM results for interior compartments 2 through 20 and the analytical solution is shown in Figure 3.4a. Figure 3.4b shows a close-up view near the lower boundary condition. That each boundary condition assumption results in different downstream profiles is evident.

It can also be seen that the differences among the three lower boundary conditions are relatively local only, i.e., they affect only the results in the lower six or seven reaches (compartments). What is interesting about this is that all of these lower boundary conditions impose conditions that are *inconsistent* with the true (analytical) solution. The correct lower boundary condition to use is, of course, the fixed value given by the analytical solution (0.47 g/m³) at a distance corresponding to compartment 21, which we are pretending we do not know.

Given that none of our boundary conditions is consistent with reality, one might then wonder how the underlying mathematics will try to resolve these inconsistencies. Will the solution satisfy the lower boundary conditions, but at the expense of being in some (perhaps significant) error throughout the entire distance downstream? We clearly see that this is not the case, even for the results for the 3 g/m³ fixed concentration assumption which is obviously not a good one. Indeed, the GEM results over all but the lowest reaches are essentially identical to the analytical solution. Thus, the underlying mathematics have not given up a great deal in accuracy in the face of these inconsistent assumptions.

The three different lower boundary conditions exhibit some notable differences. For the zero gradient condition, the upturn in concentration necessary to achieve that result is clearly seen. For the fixed boundary condition, the dramatic upturn necessary to achieve that result is also obvious. Finally, relative to the analytical results, the linear gradient profile in the lower reaches must "straighten out" somewhat, i.e., become increasingly linear to achieve that condition.

REFERENCE

Hoffman, J.D. 1992. *Numerical Methods for Engineers and Scientists.* New York: McGraw-Hill.

4 Source and Sink Terms

For purposes of the GEM, a *source* or *sink* is a gain or loss of constituent mass to or from a compartment from some mechanism *other than* the transport processes previously presented in Chapter 3. Sources and sinks are sometimes termed *fate* processes in environmental modeling. This chapter introduces fate processes into the mass balance equations presented in Chapter 3. Section 4.1 introduces several different types of linear source and sink terms. Section 4.2 shows the incorporation of those linear terms into mass balance equations. A generic representation of nonlinear source and sink terms is then introduced into the mass balance equations in Section 4.3.

4.1   LINEAR SOURCE AND SINK TERMS

Linear sources or sinks are either zero or first-order processes.[*] The GEM includes five types of linear sources and sinks: a zero-order exogenous sink, (2) a first-order exogenous sink, (3) a first-order cross-compartment source or sink, (4) a first-order cross-chemical source or sink, and (5) a first-order retardation sink that is commonly used in groundwater modeling. These are described below.

4.1.1   ZERO-ORDER EXOGENOUS SINK

This represents a loss of mass from a compartment that does not represent a gain to another (modeled) compartment or another (modeled) chemical (hence the term "exogenous") and is not a function of chemical concentration. An exogenous sink is analogous to the external loading term (W_i) previously introduced for external sources. Indeed, the zero-order exogenous sink is simply a negative loading and is specified by the user in the GEM's forcing function (**b** vector entered in the Loads. csv file) just as a loading is entered but with a negative value.

EXAMPLE 4.1

Zero-Order Sink Example

Consider a model of total suspended solids (TSS) in a completely-mixed lake. The lake consists of the water compartment (compartment 1) and a sediment compartment (compartment 2). Compartment 3 is the upstream (inflow) boundary compartment. Compartment 4 is the downstream dummy compartment. We use

---

[*] The order of a chemical process refers to the power to which the chemical concentration is raised. A zero-order process raises the concentration to the 0th power (1) so the process is a constant. A first-order process involves a power of 1.

a backward difference ($\alpha = 1$) and assume the water content is 1.0. The mass balance equation for TSS in compartment 1 is

$$\frac{d(V_1 TSS_1)}{dt} = Q_{3,1} TSS_3 - Q_{1,4} TSS_1 - v_s A_{1,2} TSS_1 + v_r A_{1,2} TSS_2 \qquad (4.1)$$

where
  V_i = volume of compartment i (L^3)
  $Q_{i,j}$ = advective flow between compartments i and j (L^3/T)
  $A_{i,j}$ = area between compartments i and j (L^2)
  v_s = settling rate (L/T)
  v_r = sediment resuspension rate (L/T).

The mass balance equation for TSS in compartment 2 is

$$\frac{d(V_2 TSS_2)}{dt} = v_s A_{1,2} TSS_1 - v_r A_{1,2} TSS_2 - v_b A_{1,2} TSS_2 = 0 \qquad (4.2)$$

where v_b = sediment burial rate (L/T).

Under the usual assumptions that neither the solids concentrations in the sediment (TSS_2) nor the sediment volume is changing with time, the time derivative in Equation (4.2) becomes zero. (The solids concentration in the sediment is simply the bulk density of the sediments themselves.) Under these assumptions, Equation (4.2) can be used to express the resuspension rate as a function of the settling and burial terms (both of which are easier to estimate than the resuspension rate), or

$$v_r = v_s \frac{TSS_1}{TSS_2} - v_b \qquad (4.3)$$

Substituting Equation (4.3) into Equation (4.1) for v_r results in

$$\frac{d(V_1 TSS_1)}{dt} = Q_{3,1} TSS_3 - Q_{1,4} TSS_1 - v_b A_{1,2} TSS_2 \qquad (4.4)$$

Thus, under our assumptions, the three-compartment model simplifies to a two-compartment model [Equations (4.1) and (4.4)], and the TSS time distribution in compartment 1 is determined simply by the boundary condition and the burial rate. The burial term ($v_b A_{2,3} TSS_2$) in Equation (4.1) involves presumably known quantities and is simply a constant negative loading, i.e., a zero-order exogenous sink. (Alternatively, as discussed below, we could model the burial term by adding an additional (dummy) compartment below the sediment compartment and model the burial sink as a cross-compartment transfer.)

4.1.2 First-Order Exogenous Sources and Sinks

These also represent a gain or loss of mass from a compartment that does not represent a loss or gain either to another modeled compartment or another modeled chemical. However, the magnitude of the source or sink is now a function of the chemical concentration. Examples of sinks are biochemical or radioactive decay, volatilization, or sediment burial—all cases where the chemical simply "goes away" from the perspective of the compartments and constituents modeled. An example of a first-order source is bacterial growth.

Exogenous first-order sources and sinks are assumed by the GEM to be volume-based. Volume-based exogenous sources and sinks are represented by the following loss term written for arbitrary chemical k in arbitrary compartment i

$$\pm k_{i,k} V_i \theta_i c_{i,k} \phi_i^{(T_i-20)} \tag{4.5}$$

where $k_{i,k}$ denotes the (user-specified) first-order exogenous source or sink rate constant with units of $1/T$, V_i is the (total) compartment volume (L^3), as previously defined, θ_i is again the water content, ϕ_i is a user-specified temperature correction factor normalized to $20°C$ (1.024 is commonly used), and T_i is the user-specified temperature of compartment i (degrees Celsius).

Note that more than one exogenous source or sink term may be used in a compartment, i.e., different processes may cause simultaneous exogenous gains and/or losses. Equation (4.5) above assumes only one for illustrative purposes but the GEM can accommodate as many as desired. Alternatively, and more simply, the k rate term could represent the aggregate rate constant across the several processes.

It should be noted that kinetic processes when simulated with numerical approaches using relatively large time steps can lead to erroneous results as demonstrated in Box 4.1.

BOX 4.1   TEMPORAL DISCRETIZATION ERROR FOR FIRST-ORDER PROCESSES

The numerical modeler (GEM or otherwise) needs to be aware of potential discretization error when simulating a dynamic process involving "fast" first-order kinetics. For example, Figures 4.1 and 4.2 show the results of GEM simulations of first-order growth of a substance (e.g., bacteria) in a batch reactor (no inflow or outflow) starting from an initial concentration of 10 g/m^3, as compared to the true (analytical) solution.

In Figure 4.1, the growth constant is assumed to be relatively slow at 0.1/day while a much faster 0.5/day rate is used in Figure 4.2. For the 0.1/day scenario, we ran a forward time (FT) Euler-type dynamic simulation (discussed in Chapter 6) using a time step of 1 day. (In fact, this example is also the most simple compartment model possible with the GEM. Two compartments were used: the batch reactor compartment and an adjacent dummy. The minimum number of compartments possible with the GEM is two, hence the dummy compartment.) The simulated results in Figure 4.1 are not perfect, but are reasonably close to the analytical results. For the 0.5/day scenario, it can be seen that that same time step is inaccurate. To achieve accuracy, we have to decrease the time step by a factor of 100, to 0.01/day.

Why is this? The problem is the temporal discretization error introduced by the numerical approximation. The analytical solution is based on the differential equation

$$\frac{dc}{dt} = kt$$

which has the solution

$$c(t) = c(0)e^{kt} \tag{4.6}$$

where $c(0)$ is the initial condition, 10 g/m³. The corresponding GEM differential equation is

$$\frac{d(Vc)}{dt} = kV\theta\phi^{T-20}$$

For constant volume, $\theta = \phi = 1$ and temperature = 20°C, and using the FT dynamic solution option, the GEM recursion equation is

$$c^{t+1} = c^t + \Delta t(kc^t) \tag{4.7}$$

where Δt is the time step and $t = 0$ corresponds to the initial condition. To use Equation (4.7), you simply start with the initial condition, 10 g/m³ and "recursively" march the solution forward in time.

You can see that Equation (4.7) is a linear approximation to the true equation (4.6). If Equation (4.6) is a highly nonlinear function (such as the analytical solution in Figure 4.2 for $k = 0.5$), you have to use very small time steps to not venture very far off the true curve as you march along it. The errors introduced from these larger time steps are not due to numerical instabilities or numerical dispersion (both are discussed later) but rather are simply discretization errors resulting from approximating a nonlinear function with a piece-wise linear approximation.

4.1.3 Cross-Compartment First-Order Sources and Sinks

We assume that all sources or sinks of mass across modeled compartment boundaries occur as *area-based* fluxes. (We are unable to conceive an example of a volume-based, cross-compartment transfer. If our imagination is discovered to be too limited, the user can specify a transfer process through the GEM's user-defined, nonlinear **f(c)** function described later in this chapter. However, such a source or sink would be treated as a nonlinear term by the GEM even if it is in fact linear. The good news is that, if linear, the iterative, quasi-Newton solution algorithm described in Chapter 5

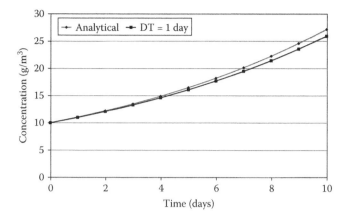

FIGURE 4.1   The 0.1/day growth constant.

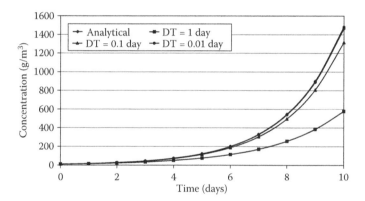

FIGURE 4.2   The 0.5/day growth constant.

should converge in a single iteration.) For sinks, these are represented relative to compartment i as

$$-c_{i,k}\theta_i \sum_j A_{i,j} v_{i,j,k} \tag{4.8}$$

where the j index denotes the summation over compartment i's neighboring compartments, $A_{i,j}$ is the surface area between compartments i and j across which the flux is occurring (L^2) and $v_{i,j,k}$ is a (possibly chemical-specific) user-supplied velocity (L/T) between compartments i and j for chemical k, sometimes referred to as a mass transfer coefficient. The water content parameter is used here to modify the interfacial area through which the flux occurs in porous media applications.

For cross-compartment area-based sources, the fluxes are from the various neighboring compartment j to compartment i, and the source term is also a summation over i's neighbors

$$+\sum_{j} v_{i,j,k} A_{i,j} \theta_j c_{j,k} \tag{4.9}$$

4.1.4 Cross-Chemical First-Order Sources and Sinks

Cross-chemical sources and sinks are chemical reactions between constituents. We assume that these are all volume-based and occur only within a given compartment. (We also cannot conceive of a simultaneous cross-chemical and cross-compartment scenario, i.e., where a chemical interacts with another chemical, but only as part of a cross-compartment process; again, the user can override this possible limitation with the $\underline{\mathbf{f}(\mathbf{c})}$ function.)

Cross-chemical sources or sinks of a given chemical k in compartment i are represented as

$$\pm V_i \theta_i \phi_i^{(T_i-20)} \sum_{l} c_{i,l} k_{k,l} \tag{4.10}$$

where the l index is a summation over chemicals and $k_{k,l}$ is now a first-order rate constant $(1/T)$ that typically (but not always, see following paragraph) represents conversion *from* chemical k *to* chemical l. If the user-specified $k_{k,l}$ is negative, the corresponding term is a sink for chemical k. If positive, it is a source. As noted in Chapter 2, in the VolumeSrcSnks.csv discussion, a loss from chemical k that is a gain to chemical l (or vice-versa) must be explicitly specified for both chemicals k and l. The GEM does not automatically provide this functionality.

While the requirement that the user must specify both, e.g., a loss from chemical k (to l) and the gain from l to k, may seem unnecessary, in fact this requirement—and the general form of (4.10) above—allows a great deal of flexibility in specifying volume-based sources and sinks. For example, we may want to specify a sink of chemical k that is *not* a function of its own concentration (e.g., $k_{i,k}c_{i,k}$) but rather a function of some other chemical's concentration ($k_{i,l}c_{i,l}$). Such would be the case, for example, if the concentration of chemical k varies stoichiometrically as a consequence of chemical l's concentration. Thus, chemical k varies directly in some manner with chemical l, but mass transfers are not occurring *between* k and l.

4.1.5 Retardation Sink for Porous Media

The so-called retardation sink presented here is often used in modeling fate and transport of sorbing chemicals in groundwater systems. *Retardation* refers to the fact that dissolved chemical in groundwater systems is often the chemical state variable of choice because only the dissolved fraction moves with the groundwater flow. The sorbed fraction is left behind, i.e., sorbed onto the immobile solids, so that the total chemical mass transported through the system is initially retarded to some extent by this sorptive loss. (Once the solids' sorptive capacity is exhausted, retardation no longer occurs.) The inclusion of the retardation sink results in a sink term unlike the three presented above in that the retardation sink becomes a modification to the left side (temporal derivative term) of the system of mass balance equations.

Although it is useful to include this retardation sink-type modification (which results in inclusion of an additional matrix $\mathbf{R}$ on the left side) only insofar as it enables the GEM to be applied to modeling dissolved chemical transport in groundwater systems, it is also possible that the $\mathbf{R}$ matrix may have application to other environmental modeling scenarios quite apart from dissolved chemical in groundwater. (The user could set the $\mathbf{R}$ elements to be anything he or she likes given the modeling context, not necessarily the parameterization that follows below.) In any event, including the $\mathbf{R}$ matrix in the generalized GEM equations and solution techniques only adds flexibility and does not impose additional restrictions, because $\mathbf{R}$ can always be trivially set to the identity matrix (1s on the main diagonal and 0's elsewhere) to effectively remove it from the system of equations.

Assume that we are modeling the dissolved fraction of a chemical that is subject to an irreversible loss due to linear equilibrium-controlled sorption to solids. Thus, the chemical in compartment i c_i is dissolved chemical (M/L$^3_{water}$) and the sorption kinetics are sufficiently fast that the equilibrium relationship between dissolved and sorbed concentrations is instantaneously reached. Finally, because we are currently considering linear systems, it is further assumed that the relationship between dissolved and sorbed concentration is linear. As discussed elsewhere (see, for example, Fetter, 1999), the sink of dissolved chemical k due to sorption is generally expressed (and modified here for our compartment i) as

$$-BD_i V_i \frac{d\bar{c}_{i,k}}{dt} \tag{4.11}$$

where

$\bar{c}_{i,k}$ = sorbed concentration with units of chemical mass per solids mass (Mc/Ms)
BD_i = solids bulk density (Ms/L$^3_{total}$)

The sink term in (4.11) is simply stating that the time rate of change (loss) in dissolved chemical mass due to sorption is equal to the rate at which the sorbed mass fraction is gaining, and multiplied by the bulk density to account for the difference in concentration metrics. Let us be completely general and assume that the bulk density and volume are time-variable so that (4.11) becomes

$$-\frac{d(BD_i V_i \bar{c}_{i,k})}{dt} \tag{4.12}$$

Under the instantaneous equilibrium assumption, the dissolved and sorbed concentrations are related in accordance with their experimental isotherms.* Under the linear assumption, the relationship is expressed as

$$\overline{c}_{i,k} = K_{d_{i,k}} c_{i,k} \tag{4.13}$$

where $K_{d_{i,k}}$ is the so-called partition coefficient (slope of the linear isotherm) for chemical k with units of L^3_{water}/Ms.

Equation (4.13) allows us to make a change in variables in the sink term (4.12), i.e., substituting from (4.13) for the sorbed concentration to result in the sink term expressed in terms of dissolved concentration as

$$-\frac{d(BD_iV_iK_{d_{i,k}}c_{i,k})}{dt} \tag{4.14}$$

Now we can write the mass balance for the mass of dissolved chemical k in compartment i (Equation 3.16), including the above retardation sink, as

$$d\frac{(V_i\theta_i c_{i,k})}{dt} = c_{i,k}\sum_j[\alpha_{ij}Q_{ij}I_{ij} + (1-\alpha_{ij})Q_{ij}(1-I_{ij}) - \theta_i E'_{ij}]$$

$$+ \sum_j c_{j,k}[(1-\alpha_{ij})Q_{ij}I_{ij} + \alpha_{ij}Q_{ij}(1-I_{ij}) + \theta_i E'_{ij}] \tag{4.15}$$

$$+ W_{i,k} \pm k_{i,k}V_i\theta_i c_{i,k}\phi_i^{(T_i-20)} - \frac{d(BD_iV_iK_{d_i}c_{i,k})}{dt}$$

where we have also included a volume-based first-order source/sink. Rearranging, we have

$$d\frac{(V_i\theta_i c_{i,k} + BD_iV_iK_{d_i}c_{i,k})}{dt} = c_{i,k}\sum_j[\alpha_{ij}Q_{ij}I_{ij} + (1-\alpha_{ij})Q_{ij}(1-I_{ij}) - \theta_i E'_{ij}]$$

$$+ \sum_j c_{j,k}[(1-\alpha_{ij})Q_{ij}I_{ij} + \alpha_{ij}Q_{ij}(1-I_{ij}) + \theta_i E'_{ij}] \tag{4.16}$$

$$+ W_{i,k} \pm k_{i,k}V_i\theta_i c_{i,k}\phi_i^{(T_i-20)}$$

Dividing by θ_i, we can express this as

---

* An isotherm is a plot of observed sorbed concentration versus dissolved concentration at equilibrium. These relationships can be linear or nonlinear.

$$d\frac{(R_{i,k}V_ic_{i,k})}{dt} = c_{i,k}\sum_j[\alpha_{ij}\frac{Q_{ij}}{\theta_i}I_{ij}+(1-\alpha_{ij})\frac{Q_{ij}}{\theta_i}(1-I_{ij})-E'_{ij}]$$

$$+\sum_j c_{j,k}[(1-\alpha_{ij})\frac{Q_{ij}}{\theta_i}I_{ij}+\alpha_{ij}\frac{Q_{ij}}{\theta_i}(1-I_{ij})+E'_{ij}] \qquad (4.17)$$

$$+\frac{W_{i,k}}{\theta_i}\pm k_{i,k}V_ic_{i,k}\phi_i^{(T_i-20)}$$

where R is the dimensionless retardation factor and is

$$R_{i,k}=1+\frac{BD_iK_{d_{i,k}}}{\theta_i} \qquad (4.18)$$

Going forward, Equation (4.17) with R defined as in (4.18) or user-specified as appropriate now constitutes the most general form of the GEM's mass balance equation for compartment i, with the understanding that the first-order source/sink term is simply a place holder for source/sink terms in general.

A final comment on the first-order source or sink term is in order. The implicit assumption of the $\pm k_{i,k}V_ic_{i,k}\varphi_i^{(T_i-20)}$ term in Equation (4.17) is that only the aqueous phase of the chemical is undergoing first-order kinetics. (Recall that c is now defined as the chemical mass concentration per unit of *water*, not *total* volume.) In the case that the total (aqueous + sorbed) concentration is undergoing the kinetics, it is appropriate (Schnoor, 1996) to increase the mass subject to the kinetics by the retardation factor* so that the source/sink term is $\pm k_{i,k}R_{i,k}V_ic_{i,k}\varphi_i^{(T_i-20)}$. This same caveat would apply to any other source/sink terms, e.g., area-based or cross-state-variable, as well.

EXAMPLE 4.2

Nonunity Water Content and Retardation

We conclude this discussion of porous media with an example to illustrate the effect of including solid material and retardation. Consider a column of soil that has a horizontal footprint 100 m wide by 100 m long and 10 m deep in which we wish to simulate the downward propagation of a conservative chemical due to vertical infiltration. We compartmentalize this soil column by letting it consist of 100 vertical layers, each 0.1 m thick. The uppermost (compartment 1) and lowermost (compartment 100) compartments are boundary/dummy compartments; therefore, we are simulating concentrations in the intermediate 98 compartments (sequentially numbered). Compartment 1 has a fixed boundary condition of 1.0 g/m³ that we will propagate downward through the column by advective flow

---

* The GEM source code includes a commented-out line in Subroutine *LoadALinearSrcSnk* that will effect this change for volume-based first-order kinetics. The user would need to un-comment this line and comment-out the current line.

under several different scenarios. Compartment 100 is a dummy and simply gives the flow leaving compartment 99 somewhere to go. Infiltrating flow is downward only and constant at 1,000 m^3/day.

So that the effects of solid material and retardation are more clearly shown, we (unrealistically) assume that the soil column is not subject to vertical dispersion; therefore the only transport mechanism is vertical advection. We also assume that the chemical is conservative, except for sorption to the soil particles. Because the flow and upgradient boundary concentration are constant, and the chemical is kinetically conservative, the steady state concentration in the soil pore water throughout the column is the same as the boundary condition, 1.0 g/m^3.

We consider three scenarios. Scenario 1 is a benchmark for Scenarios 2 and 3 and assumes a water content of 1.0 (θ =1) and no partitioning (K_d = 0). (Scenario 1 is simply a column filled with water that is moving slowly downward.) The constant flow velocity for Scenario 1 is 1,000/10,000 = 0.1 m/day. Although we assume no physical dispersion, numerical dispersion is always an issue (see Sections 4.4 and 4.5), and we would also like to have zero numerical dispersion for purposes of these examples.

One advantage of the Euler dynamic method is that it can be set up to result in zero numerical dispersion by a judicious selection of the time step and spatial differencing parameter α. As described in Section 4.5, we chose a backward difference ($\alpha = 1$) and a time step equal to the time that the flow takes to traverse an individual 0.1 m thick compartment or 1 day. We assume that the chemical concentration throughout the column is 0 before the simulation begins. These represent the initial condition.

Thus, for our conservative chemical with no dispersion (physical or numerical) and an upgradient boundary condition of 1.0 g/m^3, we would expect to see the 1.0 g/m^3 "front" of the chemical concentration vertical profile advance down the column in 1-day increments in scenario 1. After the first day, compartment 2 (first modeled compartment) will reach 1.0 g/m^3. After day 2, compartment 3 (second modeled compartment) will have reached 1.0, and so on. After, say, day 20, compartments 2 through 21 will all be at 1.0 g/m^3 while all compartments below are still at 0 g/m^3.

Running the GEM using the Euler option as described above (backward difference and 1-day time step) results in precisely this behavior. Figure 4.3 includes a snapshot of the vertical profile at day 20, and indeed the front has advanced to compartment 21, while all lower compartments remain at 0.

Scenario 2 is identical to Scenario 1 except we now assume a water content of 0.5 (constant throughout the column and temporally), i.e., 50% of the soil column is non-water, which we assume in this case is solids (it could also be airspace). As one would expect, the effect of this change should be to reduce the volume of each compartment through which water can flow by 50%, which in turn will double the vertical flow velocity to 0.2 m/day. Thus, the front should propagate downward at twice the rate of Scenario 1.

We ran the GEM under this scenario again using the Euler option and a backward difference, but now using a time step of 0.5 days (to maintain both stability and zero numerical dispersion). Figure 4.3 also shows a snapshot of the vertical profile at 20 days for this scenario. As expected, the front has now reached compartment 41, twice the depth attained with the unity water content of Scenario 1. (The same result is also obtained by running the GEM with a unity water content and the volume of each compartment reduced by one-half.) The relatively simple effect of a non-unity water content should now be clear.

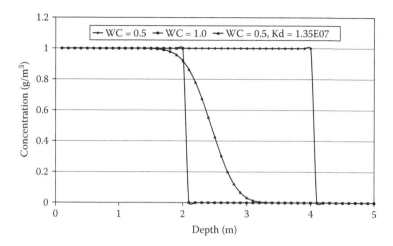

FIGURE 4.3 Porous media example.

Scenario 3 adds linear partitioning to Scenario 2. We assume now that the bulk density of the soil column solids is 2.5×10^6 g/m³ and the partition coefficient (K_d) is 1.35×10^{-7} m³/g. This K_d corresponds to that of benzene, an organic chemical that is mildly sorbing. We again ran this using the Euler option with a backward difference and a 0.5-day time step. Figure 4.3 also shows a 20-day snapshot of the vertical profile for Scenario 3. By comparison to Scenario 2, it can be seen that the 1.0 g/m³ front has been significantly retarded due to partitioning and the front is now somewhat spread out vertically as opposed to the absolutely sharp front of Scenario 2. (The concentration in compartment 41 is very small, but non-zero, while compartments 42 through 99 are identically zero because the front has not reached them yet.)

It should be noted that the steady-state concentrations with partitioning will still be 1.0 g/m³ in all compartments, as shown by the concentrations near the top of the soil column in Scenario 3 that reached steady state after 20 days. Thus, partitioning is a transitory phenomenon only before steady state is reached. This may seem counterintuitive but can be understood mathematically by examining the governing system of differential equations [Equations (4.1) through (4.10)]. The retardation matrix **R** only appears in the equations within the time derivative. At steady state, this time derivative is zero so that **R** simply falls out of the system of equations.

4.2 SYSTEM OF FATE AND TRANSPORT, MASS BALANCE EQUATIONS FOR MULTIPLE CHEMICALS

We can now include our various linear sources and sinks into the mass balance equation for chemical k in compartment i. Recall from Equation (3.16) that the coefficients of $c_{i,k}$ were grouped for convenience. All (linear) sinks of chemical k in compartment i also go into the group of coefficients that modify $c_{i,k}$. All sources (from other compartments or other chemicals) enter the other right side terms modifying $c_{j,k}$. The complete fate and transport mass balance equation for chemical k in compartment i is then

$$
d\frac{(R_{i,k}V_i c_{i,k})}{dt} = c_{i,k}\left[\alpha\sum_j \frac{Q_{ij}}{\theta_i}I_{ij} + (1-\alpha)\sum_j \frac{Q_{ij}}{\theta_i}(1-I_{ij})\right.
$$

$$
\left. -\sum_j E'_{ij} - k_{i,k}V_i \ -\sum_j A_{i,j}v_{i,j,k} - V_i\sum_l k_{k,l}\right]
$$

$$
+(1-\alpha)\sum_j \frac{Q_{ij}}{\theta_i}I_{ij}c_{j,k} + \alpha\sum_j \frac{Q_{ij}}{\theta_i}(1-I_{ij})c_{j,k} + \sum_j E'_{ij}c_{j,k}
$$

$$
+\frac{W_{i,k}}{\theta_i} + \sum_j v_{i,j,k}A_{i,j}c_{j,k} + V_i\sum_l c_{i,l}k_{l,k}\varphi_i^{T_i-20}
$$

(4.19)

where R_i is given by Equation (4.18). Writing Equation (4.19) for all NIC interior compartments and all MaxNSV chemicals, we can express the system of NEQN equations in matrix notation as

$$
\frac{d(\mathbf{RV}\underline{c})}{dt} = \mathbf{A}\underline{c} + \underline{b}
$$

(4.20)

where $\mathbf{R}$ is now the NEQN × NEQN main diagonal matrix replacing Θ in Equation (3.17)—it is more general—with NIC-repeating main diagonal elements given by Equation (4.18).

The $\mathbf{A}$ matrix of Equation (4.20), now including transport, multiple chemicals, and all types of linear sources and sinks, is shown in Figure 4.4. It no longer necessarily has zero off-main diagonal sub-matrices (as shown previously for transport only in Figure 3.2) because of the potential cross-chemical coupling terms. Within a main diagonal sub-matrix, i.e., the $\mathbf{A_{ii}}$ sub-matrices, the main diagonal elements include all sinks (exogenous, cross-compartment, and cross-chemical). The off-diagonal elements of a main diagonal sub-matrix include sources between compartments within a given chemical, e.g., settling or resuspension from one compartment to another. The off-main diagonal sub-matrices, i.e., the $\mathbf{A_{ij}}$ sub-matrices (i not equal to j), are main diagonal matrices only. All off-diagonal elements are zero. The main diagonals of these off-main diagonal sub-matrices are the cross-chemical source terms. Note that a *non-zero* off-main diagonal element of these sub-matrices would represent a cross-compartment cross-chemical sink that we assume does not exist.

$$
\begin{bmatrix}
(A_{1,1}) & (A_{1,2}) & \cdots & & (A_{1,\text{MaxNSV}}) \\
(A_{2,1}) & (A_{2,2}) & (A_{2,3}) & \cdots & (A_{2,\text{MaxNSV}}) \\
 & \cdots & (A_{1,1}) & \cdots & \vdots \\
(A_{\text{MaxNSV}}) & & \cdots & & (A_{\text{MaxNSV,MaxNSV}})
\end{bmatrix}
\begin{bmatrix}
(\underline{c}_1) \\
(\underline{c}_2) \\
\vdots \\
(\underline{c}_{\text{MaxNSV}})
\end{bmatrix}
+
\begin{bmatrix}
(\underline{b}_1) \\
(\underline{b}_2) \\
\vdots \\
(\underline{b}_{\text{MaxNSV}})
\end{bmatrix}
$$

$\qquad\qquad\qquad$ A $\qquad\qquad$ $\underline{c}$ $\qquad\qquad$ $\underline{b}$

FIGURE 4.4 Matrix structure of system of mass balance equations for multiple chemicals.

EXAMPLE 4.3

Multiple Chemicals

An example problem illustrates the system of equations and matrix structure for multiple chemicals. The example is for two chemicals in a three-compartment (3 nonboundary/nondummy compartments) system. The example is again a surface water quality problem, this time for phosphorus and algal uptake/release of phosphorus in a vertically-stratified lake during the algae growing season. Algae take up PO_4 during growth as a nutrient, and release it back to the PO_4 pool through death and subsequent mineralization. This example is a highly simplified version of this very complex ecological process. The water content parameter Θ_i is assumed to be 1 for all compartments and sorption is not considered. (Thus, the **R** matrix is identically the identity matrix **I**.)

Chemical 1 is the dissolved phosphorus concentration that is available for algal uptake, i.e., PO_4. Chemical 2 represents the phosphorus mass tied up in algal cells. That is, chemical 2 is the concentration of algal phosphorus in the water, not the concentration of the algae themselves. Hereafter, we will denote chemical 1 as *NAP* (non-algal phosphorus) and chemical 2 as *AP* (algal phosphorus).

The environmental medium modeled is a stratified lake consisting of an upper water column (epilimnion) as compartment 1, a lower water column (hypolimnion) as compartment 2, and a sediment layer as compartment 3. All three are interior compartments within which the two chemicals (*NAP* and *AP*) are simulated. In addition, there are two exterior compartments. Compartment 4 is a boundary exterior compartment that represents the inflowing stream upstream of the modeled lake. Compartment 5 is a dummy exterior compartment located downstream of the modeled lake that simply serves as a receptacle for the lake's outflowing constituent mass.[*] The five compartments are illustrated in Figure 4.5. An external loading of *NAP* (WNAP) is assumed to occur to the epilimnion. The mass balance equations are developed below by compartment.

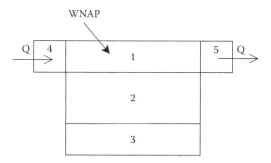

FIGURE 4.5   Compartments for example problem.

COMPARTMENT 1 (EPILIMNION)

We assume that transport is due to advective flow only from compartment 4 to compartment 1, and from compartment 1 to compartment 5. Transport between compartments 1 and 2 is due to dispersion only. *NAP* is assumed to be completely dissolved and does not settle. *AP* is assumed to be particulate and settles. Temperature effects on kinetics are ignored.

For *NAP* in compartment 1, the sources are the external loading and an internal source due to release of phosphorus from the *AP*. The upstream boundary condition is assumed to be 0 mg/L. The sink is first-order uptake by algae. We also assume that $\alpha = 1$, i.e., a backward difference is used (an assumption consistent with the physical scenario of no dispersion, i.e., there is no physical mechanism by which a downstream concentration can affect an upstream concentration). Under these assumptions, the mass balance equation for *NAP* in compartment 1 (NAP_1) is

$$\frac{d(V_1 NAP_1)}{dt} = Q_{4,1} NAP_4 - Q_{1,5} NAP_1 + \frac{E_{1,2} A_{1,2}}{L_{1,2}} (NAP_2 - NAP_1)$$

$$- k_{1,2} V_1 NAP_1 + k_{2,1} V_1 AP_1 + W_{1,1} \tag{4.21a}$$

where
$k_{1,2}$ = uptake rate from *NAP* to *AP* (1/day)
$k_{2,1}$ = release rate from *AP* to *NAP* (1/day)
V_1 = volume of compartment 1 (m³)
$E_{1,2}$ = dispersion coefficient between compartments 1 and 2 (m²/day)
$A_{1,2}$ = area between compartments 1 and 2 (m²)
$L_{1,2}$ = distance between centroids of compartments 1 and 2 (m)
$Q_{4,1}$ = advective flow rate (m³/day) from compartment 4 to compartment 1
$Q_{1,5}$ = advective flow rate (m3/day) from compartment 1 to compartment 5
NAP_4 = upstream boundary condition (g/m³)
$W_{1,1}$ = external *NAP* loading to compartment 1 (g/day)

Note that we are not assuming that the volumes are time-variable, but we have left them inside the derivative term for generality and consistency with the previous mathematical development. Similarly, for *AP* in Compartment 1 (AP_1), the mass balance equation is

$$\frac{d(V_1 AP_1)}{dt} = Q_{4,1} AP_4 - Q_{1,5} AP_1 + \frac{E_{1,2} A_{1,2}}{L_{1,2}} (AP_2 - AP_1)$$

$$- k_{2,1} V_1 AP_1 + k_{1,2} V_1 NAP_1 - v_{1,2} A_{1,2} AP_1 \tag{4.21b}$$

where $v_{1,2}$ = settling velocity between compartments 1 and 2 (m/day).

COMPARTMENT 2 (HYPOLIMNION)

In the hypolimnion, we assume that algae do not grow due to light limitations, but only die and release phosphorus back to the *NAP* pool or settle into the sediments. Dispersion of both *AP* and *NAP* again occurs between compartments 1 and 2. Between compartments 2 and 3 (sediment), we assume that molecular diffusion occurs and acts only on the (dissolved) *NAP*. There is no advective transport for compartment 2. Settling of AP from the epilimnion is a source for AP in the hypolimnion, while settling to the sediment is a sink. With these assumptions, the mass balance equations are

$$\frac{d(V_2 NAP_2)}{dt} = \frac{E_{1,2}A_{1,2}}{L_{1,2}}(NAP_1 - NAP_2) + \frac{E_{2,3}A_{2,3}}{L_{2,3}}(NAP_3 - NAP_2) + k_{2,1}V_2 AP_2 \quad (4.21c)$$

$$\frac{d(V_2 AP_2)}{dt} = \frac{E_{1,2}A_{1,2}}{L_{1,2}}(AP_1 - AP_2) - k_{2,1}V_2 AP_2 + v_{1,2}A_{1,2}AP_1 - v_{2,3}A_{2,3}AP_2 \quad (4.21d)$$

where
  $E_{2,3}$ = diffusion coefficient (m²/day) between compartments 2 and 3
  $A_{2,3}$ = area (m²) between compartments 2 and 3
  $L_{2,3}$ = distance between centroids of compartments 2 and 3 (m)
  V_2 = volume of compartment 2 (m³)
  $v_{1,2}$ = settling velocity between compartments 1 and 2 (m/day)
  $v_{2,3}$ = settling velocity between compartments 2 and 3 (m/day)

COMPARTMENT 3 (SEDIMENTS)

Compartment 3 is essentially a sink of settled algal phosphorus because no resuspension back into the water column is assumed. The settled algae release their phosphorus back into the NAP pool in compartment 3 that can then be diffused back into compartment 2 (and subsequently to compartment 1), so there is a feedback mechanism for NAP from the sediments to the water column compartments. These are the only mechanisms assumed for compartment 3. Accordingly, the mass balance equations are

$$\frac{d(V_3 NAP_3)}{dt} = \frac{E_{2,3}A_{2,3}}{L_{2,3}}(NAP_2 - NAP_3) + k_{2,1}V_3 AP_3 \quad (4.21e)$$

where V_3 = volume of compartment 3 (m³).

$$\frac{d(V_3 AP_3)}{dt} = v_{2,3}A_{2,3}AP_2 - k_{2,1}V_3 AP_3 \quad (4.21f)$$

With respect to the generalized matrix Equation (4.20), the $\mathbf{V}$, $\mathbf{A}$, $\underline{\mathbf{c}}$, and $\underline{\mathbf{b}}$ matrices and vectors are presented in Figures 4.6 and 4.7 for this example. Again, $\mathbf{R} = \mathbf{I}$ (identity matrix) for this application. The individual elements of the $\mathbf{A}$ matrix in Figure 4.7 (scalar $a_{i,j}$ values) are:

$a_{1,1} = -Q_{1,5} - E_{1,2}A_{1,2}/L_{1,2} - k_{1,2}V_1$
$a_{1,2} = E_{1,2}A_{1,2}/L_{1,2}$
$a_{1,4} = k_{2,1}V_1$
$a_{2,1} = E_{1,2}A_{1,2}/L_{1,2}$
$a_{2,2} = -E_{1,2}A_{1,2}/L_{1,2} - E_{2,3}A_{2,3}/L_{2,3}$
$a_{2,3} = E_{2,3}A_{2,3}/L_{2,3}$
$a_{2,5} = k_{2,1}V_2$
$a_{3,2} = E_{2,3}A_{2,3}/L_{2,3}$

$$
\left[
\begin{array}{ccc|ccc}
\begin{pmatrix} V_1 & 0 & 0 \\ 0 & V_2 & 0 \\ 0 & 0 & V_3 \end{pmatrix} & & & \begin{pmatrix} 0 & 0 & 0 \\ 0 & 0 & 0 \\ 0 & 0 & 0 \end{pmatrix} \\[6pt]
\begin{pmatrix} 0 & 0 & 0 \\ 0 & 0 & 0 \\ 0 & 0 & 0 \end{pmatrix} & & & \begin{pmatrix} V_1 & 0 & 0 \\ 0 & V_2 & 0 \\ 0 & 0 & V_3 \end{pmatrix}
\end{array}
\right]
$$

FIGURE 4.6   Structure of V matrix for example.

$$
\begin{bmatrix}
\begin{pmatrix} a_{1,1} & a_{1,2} & 0 \\ a_{2,1} & a_{2,2} & a_{2,3} \\ 0 & a_{3,2} & a_{3,3} \end{pmatrix} & \begin{pmatrix} a_{1,4} & 0 & 0 \\ 0 & a_{2,5} & 0 \\ 0 & 0 & a_{3,6} \end{pmatrix} \\
\begin{pmatrix} a_{4,1} & 0 & 0 \\ 0 & 0 & 0 \\ 0 & 0 & 0 \end{pmatrix} & \begin{pmatrix} a_{4,4} & a_{4,5} & 0 \\ a_{5,4} & a_{5,5} & 0 \\ 0 & a_{6,5} & a_{6,6} \end{pmatrix}
\end{bmatrix}
\begin{bmatrix} NAP_1 \\ NAP_2 \\ NAP_3 \\ AP_1 \\ AP_2 \\ AP_3 \end{bmatrix}
+
\begin{bmatrix} Q_{4,1}NAP_4 + W_{1,1} \\ 0 \\ 0 \\ Q_{4,1}AP_4 \\ 0 \\ 0 \end{bmatrix}
$$

$$\quad\quad\quad\quad A \quad\quad\quad\quad\quad\quad\quad\quad\quad\quad\quad \underline{c} \quad\quad\quad\quad \underline{b}$$

FIGURE 4.7   Matrix structure of example problem.

$$a_{3,3} = -E_{2,3}A_{2,3}/L_{2,3}$$
$$a_{3,6} = k_{2,1}V_3$$
$$a_{4,1} = k_{1,2}V_1$$
$$a_{4,4} = -Q_{1,5} - E_{1,2}A_{1,2}/L_{1,2} - k_{2,1}V_1 - v_{1,2}A_{1,2}$$
$$a_{4,5} = E_{1,2}A_{1,2}/L_{1,2}$$
$$a_{5,4} = v_{1,2}A_{1,2} + E_{1,2}A_{1,2}/L_{1,2}$$
$$a_{5,5} = -E_{1,2}A_{1,2}/L_{1,2} - k_{2,1}V_2 - v_{2,3}A_{2,3}$$
$$a_{6,5} = v_{2,3}A_{2,3}$$
$$a_{6,6} = -k_{2,1}V_3$$

The point of this example is to demonstrate the matrix structure of the system of dynamic mass balance equations for multiple state variables. Nonetheless, it is interesting to assign numerical values to the example's parameters and loadings and run a simulation with the GEM. Therefore, we ran a dynamic simulation using a daily time step and assigned the following (time-constant) values.

$V_1 = 10,000$ m³
$V_2 = 90,000$ m³
$V_3 = 1,000$ m³
$A_{1,2} = A_{2,3} = 10,000$ m²
$Q_{4,1} = Q_{1,5} = 10,000$ m³/day
$L_{1,2} = 5$ m (depths of compartments 1 and 2 are 1 m and 9 m, respectively)
$L_{2,3} = 4.55$ m (depth of compartment 3 is 0.1 m)
$k_{1,2} = 0.5$/day
$k_{2,1} = 0.05$/day
$E_{1,2} = 1$ m²/day
$E_{2,3} = 0.1$ m²/day
$NAP_4 = AP_4 = 0$ g/m³
$v_{1,2} = v_{2,3} = 1$ m/day

We let the loading vary over time in a step-wise decreasing fashion as:

$W_{1,1}$ = 10 g/day for time 0 to day 50
$W_{1,1}$ = 5 g/day for time 51 to day 100
$W_{1,1}$ = 0 g/day for time after day 100

The results of the GEM dynamic simulation for 200 days are shown in Figures 4.8 through 4.10. We used a BTBS method discussed in Chapter 6. The response to the step-wise decrease in the time-varying loading pattern is evident. Note that it takes approximately 100 days following complete cessation of the load for

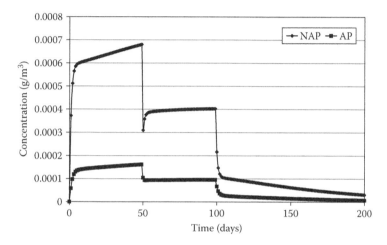

FIGURE 4.8 Concentration time series in epilimnion.

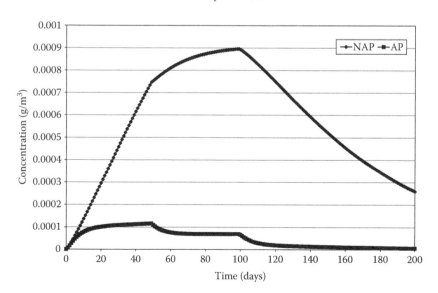

FIGURE 4.9 Concentration time series in hypolimnion.

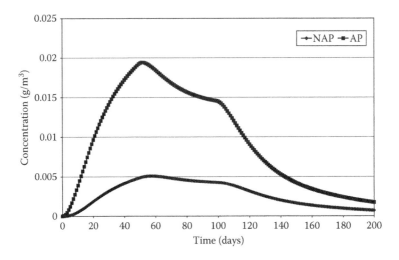

FIGURE 4.10   Concentration time series in sediments.

concentrations in the epilimnion (compartment 1) to approach zero. However, in the hypolimnion (compartment 2) and sediments (compartment 3), much more time will be required. In the epilimnion and hypolimnion, the concentrations of *NAP* exceed *AP*, as would be expected given that *NAP* has the external loading and relatively low *AP* release rate. However, the opposite is true in the sediments because they are essentially traps of *AP*.

We also note that the fairly abrupt changes in concentration in the epilimnion following changes in loading are somewhat dampened in the other compartments, as would be expected given the transport bottleneck due to only dispersive transport (and settling) across the thermocline and an even greater diffusion bottleneck across the water–sediment interface. Finally, we note that the *NAP* concentrations in compartments 2 and 3 exceed *NAP* in compartment 1, despite the compartment 1 *NAP* external loading. This is because *NAP* is relatively quickly advected from compartment 1 in the outflow, whereas there is no advective loss from compartments 2 or 3.

4.3   SOURCE AND SINK TERMS FOR NONLINEAR SYSTEMS

4.3.1   General Approach

Section 4.1 presented several different types of linear sources and sinks. It was practical to precisely specify the functional forms of those terms because the linearity assumption greatly reduces the number of possibilities of mathematical forms, and those presented are quite commonly used. However, when the linearity assumption is relaxed and one considers the many different modeling applications in which the GEM might be applied, it is inappropriate to pre-specify particular mathematical forms of nonlinear sources and sinks. Therefore, we simply present the most general nonlinear form and its generalized solution technique and leave it to the user to specify the specific mathematical structure using the GEM's shell procedure, discussed in Chapter 2 for any particular application.

We also assume that a GEM application can have both linear and nonlinear sources and sinks. Therefore, we retain the most general expression of the system of mass balance differential equation for linear systems (Equation 4.2) and add a vector of (generalized) source and sink terms to that equation to denote nonlinear sources and sinks. We also acknowledge that the **R** matrix (used for incorporating the retardation term in porous media systems) may also involve nonlinear functions of the chemicals. It is important to note that we continue to assume that transport processes remain linear.

Under this approach, we can extend Equation 4.20 to include nonlinear sources/sinks as

$$\frac{d(\mathbf{R}(\underline{c})\mathbf{V}\underline{c})}{dt} = \mathbf{A}\underline{c} + \underline{f}(\underline{c}) + \underline{b} \tag{4.22}$$

where $\underline{f}(\underline{c})$ is a generalized NEQN × 1 vector of nonlinear sources or sinks[*] and $\mathbf{R}(\underline{c})$ also denotes that elements of the **R** matrix may involve nonlinear functions of $\underline{c}$. $\mathbf{R}(\underline{c})$ and $\underline{f}(\underline{c})$ denote (unspecified) functional forms of $\underline{c}$. Their specific functional forms will be application-specific. All other terms in Equation (4.22) are as previously described. In particular, matrix **A** continues to consist of linear transport terms and (any) linear sources and sinks. Equation (4.22) is now the most generalized equation representing the GEM. (Absence of nonlinear sources or sinks simply means that all elements of $\underline{f}(\underline{c})$ are zero and **R** is not a function of $\underline{c}$.)

Two examples of using nonlinear source/sink terms follows. The first involves the nonlinearity in the $\underline{f}(\underline{c})$ term and is a surface water example while the second is a groundwater example and involves the nonlinearity in the $\mathbf{R}(\underline{c})$ retardation term.

EXAMPLE 4.4

Nonlinear Kinetics in Surface Water System

This example illustrates substrate-limited growth of bacteria in a single surface water compartment using Michaelis-Menten (sometimes called Monod) nonlinear kinetics. It is based on examples presented by Chapra (1997). For simplicity, it is assumed that organic carbon is the substrate and also the measure of bacterial biomass.

As bacteria grow, they uptake the substrate. When the concentration of the substrate is in abundance, bacteria can grow at their maximum rate. However, as the available substrate resource becomes increasingly depleted, their rate of growth decreases, eventually becoming zero or even negative. Thus, in contrast to use of a constant growth rate as we used in first-order (linear) reactions, the growth rate needs to change in response to these changing environmental conditions.

---

[*] It should be noted that *linear* sources/sinks could also be expressed in $\underline{f}(\underline{c})$. Thus, to the extent that a user has a linear source/sink that is not included in the GEM's inventory of linear functional forms (Section 4.1), these sources/sinks can be included in $\underline{f}(\underline{c})$. However, the GEM would then assume nonlinearity and use Newton's method to solve. (Very quick convergence should result however.)

For example, a first-order mass balance equation for bacterial growth in a single compartment subject to inflow and outflow is

$$\frac{d(Va)}{dt} = Qa_{in} - Qa + k_g a V \tag{4.23}$$

where
   a = bacterial TOC concentration (M/L^3)
   k_g = (constant) growth rate (T^{*1})

This formulation would result in exponential growth regardless of the availability of the substrate. In contrast, the Michaelis-Menten formulation expresses the growth rate as a nonlinear function of the substrate concentration as

$$k_g = \frac{k_{g,max}s}{k_s + s} \tag{4.24}$$

where
   s = substrate TOC concentration (M/L^3)
   $k_{g,max}$ = maximum growth rate (T^{-1})
   k_s = half-saturation constant (M/L^3)

k_s is the substrate concentration at which the growth rate is ½ of its maximum rate ($k_{g,max}$). Thus, at very high s-concentrations ($s \gg k_s$), Equation (4.24) approaches

$$k_g \approx k_{g,max}$$

i.e., a constant (exponential) rate. At low concentrations, ($s \ll k_s$), it approaches

$$k_g \approx \frac{k_{g,max}}{k_s} s$$

i.e., the rate is directly proportional to the substrate supply. Therefore, the Michaelis-Menten expression of k_g as a nonlinear function of available s allows k_g to vary between its maximum ($k_{g,max}$) rate and zero in accordance with the s concentration, as we desire. We use the Michaelis-Menten kinetics in a two-chemical (a,s) model for a single compartment. Bacterial death and return of bacterial TOC back to the dissolved (available) substrate TOC pool is also included.

The compartment scenario is shown in Figure 4.11 where a and s concentrations in compartment 2 are modeled and compartments 1 and 3 are boundary and dummy compartments, respectively.

FIGURE 4.11   Compartments for example.

A mass balance of bacterial TOC in compartment 2 is

$$\frac{d(V_2 a_2)}{dt} = Q_{1,2}a_1 - Q_{2,3}a_2 + k_g(s)V_2 a_2 - k_d V_2 a_2 - k_r V_2 a_2$$

where
k_d = death rate (T^{-1})
k_r = decay rate (T^{-1})

The death rate returns bacterial TOC to the substrate pool whereas the decay rate does not. Using the Michaelis-Menten model for $k_g(s)$, the above equation can be written as

$$\frac{d(V_2 a_2)}{dt} = Q_{1,2}a_1 - Q_{2,3}a_2 + \frac{k_{g,max} s_2}{k_s + s_2} V_2 a_2 - k_d V_2 a_2 - k_r V_2 a_2 \qquad (4.25)$$

A mass balance of substrate TOC in compartment 2 is

$$\frac{d(V_2 s_2)}{dt} = Q_{1,2}s_1 - Q_{2,3}s_2 - Y_c \frac{k_{g,max} s_2}{k_s + s_2} V_2 a_2 + k_d V_2 a_2 \qquad (4.26)$$

where Y_c = yield coefficient (M TOC substrate/M TOC bacteria). Writing Equations (4.25) and (4.26) in the GEM matrix notation we have

$$\frac{d(\mathbf{R}(\underline{c})\mathbf{V}\underline{c})}{dt} = \mathbf{A}\underline{c} + \underline{f}(\underline{c}) + \underline{b}$$

where, for this example,

$$\mathbf{R}(\underline{c}) = \begin{pmatrix} 1 & 0 \\ 0 & 1 \end{pmatrix}$$

i.e., we are ignoring any partitioning and the compartment 2 water content is unity. Temperature effects are also ignored.

$$\mathbf{V} = \begin{pmatrix} V_2 & 0 \\ 0 & V_2 \end{pmatrix}$$

$$\mathbf{A}\underline{c} = \begin{pmatrix} -Q_{2,3} - k_d V_2 - k_r V_2 & 0 \\ k_d V_2 & -Q_{2,3} \end{pmatrix} \begin{pmatrix} a_2 \\ s_2 \end{pmatrix}$$

$$\underline{f}(\underline{c}) = \begin{pmatrix} \dfrac{k_{g,max} s_2}{k_s + s_2} V_2 a_2 \\[4mm] -Y_c \dfrac{k_{g,max} s_2}{k_s + s_2} V_2 a_2 \end{pmatrix}$$

$$\underline{b} = \begin{pmatrix} -Q_{1,2}a_1 \\ -Q_{1,2}s_1 \end{pmatrix}$$

We are assuming no external loadings into compartment 2 for this example. Therefore, the **b** vector includes only boundary conditions.

We ran two different scenarios for this example. The first assumed that compartment 2 is a closed vessel with no inflow or outflow (batch reactor). The initial concentration of the substrate is 1,000 g/m³ while the initial concentration of bacteria is 1 g/m³. Although the volume of compartment 2 is irrelevant to this batch reactor scenario, the GEM requires a volume so we used 1 m³. (Any volume will give the same solution because no mass loading inputs are used and there is no inflow or outflow.) Other parameters of the example problem are

k_{gmax} = 5/day
k_s = 150 g/m³
k_d = 0.2/day
k_r = 0.1/day
Y_c = 1 g substrate/g bacteria

We first ran the batch reactor scenario using the Euler method (see Chapter 6). Although this is a nonlinear problem because of the Michaelis-Menten growth kinetics, the Euler dynamic solution method does not require solving simultaneous nonlinear equations (discussed in Chapter 6) when the matrix **R** is not a function of **c**, as is the case in this example. Using a time step of 0.1 day, the Euler solution from day 0 to day 20 is shown in Figure 4.12.

We note that the 0.1 day time step results in some numerical instability between approximately days 2 and 8. Accordingly, we halved the time step to 0.05 days and ran the Euler solution again. (Stability characteristics of the Euler method are discussed in Chapter 6. We are here simply trying to provide a brief preview of these accuracy issues.) The completely stable result is shown in Figure 4.13; thus, the Euler time step must be less than 0.1 day to ensure stability.

Regarding the solution, we see that, as intended using the nonlinear Michaelis-Menten growth formulation, bacteria growth is very rapid when the substrate is in abundance. Within the first 3 days, the bacteria concentration increases from its initial value to nearly 1,000 g/m³, which is mirrored by a concomitant decrease in the initial substrate concentration. As the substrate concentration decreases, however, the growth rate diminishes significantly and, indeed, is more than offset

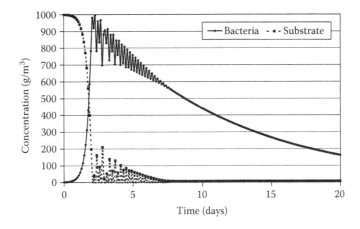

FIGURE 4.12   Results of nonlinear kinetics in batch reactor at 0.1 day time step.

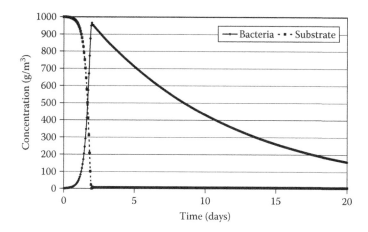

FIGURE 4.13 Results of nonlinear kinetics in batch reactor at 0.05-day time step.

by the decay rate. (With a zero decay rate, the bacteria concentration reaches its maximum value and remains at this level.)

Next, we ran the problem allowing an inflow and outflow from the compartment. The flow rate is 1 m³/day and the compartment volume was increased to 100 m³, resulting in a high hydraulic detention of 100 days. We assume an inflow boundary condition of 1,000 g/m³ for the substrate and 0 for the bacteria. The initial condition for substrate is 0 g/m³, and continues to be 1 for the bacteria. All other parameters remain identical to the batch scenario.

Running the problem using the Euler method (0.05 day time step) gives the results shown in Figure 4.14. Unlike the batch scenario in which bacterial growth began at a rapid rate due to the initially high substrate concentration, note the lag period before growth becomes exponential as the substrate concentration increases from its initial zero concentration as the substrate is transported into the compartment. As the substrate concentration peaks and begins to decline due to bacterial uptake and decay, the bacteria concentration begins to level off. After

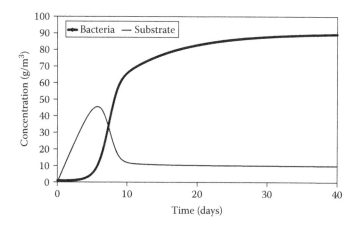

FIGURE 4.14 Results of nonlinear kinetics in flow-through reactor at 0.05-day time step.

approximately 30 days, a steady-state condition is achieved. Indeed, running the GEM in steady-state mode for this problem results in concentrations of 90.0 g/m^3 and 9.91 g/m^3 for bacteria and substrate, respectively.

EXAMPLE 4.5

Nonlinear Sorption in Groundwater System

This example is a continuation of the groundwater system example presented in Section 4.1.5. Recall that the last scenario in that example included linear partitioning of benzene in a 100-compartment soil column subject to constant vertical infiltration. Here we contrast the linear sorption results of Example 4.2 with nonlinear sorption using the Freundlich isotherm model that expresses the sorbed concentration as a nonlinear function of the dissolved concentration:

$$\bar{c}_{i,k} = K_{i,k} c_{i,k}^{N_{i,k}} \tag{4.27}$$

where $\bar{c}_{i,k}$ denotes sorbed concentration (M/M) and $K_{i,k}$ (L^3/M) and $N_{i,k}$ (unitless) are parameters. If $N_{i,k} = 1$, then $K_{i,k}$ would be the linear sorption parameter, $Kd_{i,k}$. Analogous to Equation (4.14), the sink of dissolved chemical due to sorption is

$$-\frac{d(BD_i V_i K_{i,k} c_{i,k}^{N_{i,k}})}{dt} \tag{4.28}$$

and the mass balance equation for compartment i can be written analogously to Equation (4.16) and also including the first-order sink term as an example linear source or sink as

$$d\frac{(V_i \theta_i c_{i,k} + BD_i V_i K_i c_{i,k}^{N_{i,k}})}{dt} = c_{i,k} \sum_j \left[\alpha_{ij} Q_{ij} l_{ij} + (1 - \alpha_{ij}) Q_{ij} (1 - l_{ij}) - \theta_i E'_{ij} \right]$$

$$+ \sum_j c_{j,k} \left[(1 - \alpha_{ij}) Q_{ij} l_{ij} + \alpha_{ij} Q_{ij} (1 - l_{ij}) + \theta_i E'_{ij} \right] \tag{4.29}$$

$$- k_{i,k} V_i \theta_i c_{i,k} \varphi_i^{(T_i - 20)}$$

We can factor out $c_{i,k}$ in the derivative term and again divide through by θ_i to put this in a form amenable to our retardation parameter $R_{i,k}$ as

$$d\frac{(R_{i,k} V_i c_{i,k})}{dt} = c_{i,k} \sum_j \left[\alpha_{ij} \frac{Q_{ij}}{\theta_i} l_{ij} + (1 - \alpha_{ij}) \frac{Q_{ij}}{\theta_i} (1 - l_{ij}) - E'_{ij} \right]$$

$$+ \sum_j c_{j,k} \left[(1 - \alpha_{ij}) \frac{Q_{ij}}{\theta_i} l_{ij} + \alpha_{ij} Q_{ij} (1 - l_{ij}) + E'_{ij} \right] \tag{4.30}$$

$$- k_{i,k} V_i c_{i,k} \varphi_i^{(T_i - 20)}$$

where the retardation parameter $R_{i,k}$ is now a function of the concentration (making this a nonlinear system):

$$R_{i,k} = 1 + \frac{BD_i K_{i,k} c_{i,k}^{N_{i,k}-1}}{\theta_i} \qquad (4.31)$$

The GEM was run using the Freundlich isotherm model as Scenarios 4 and 5 (to extend the linear scenarios of Section 4.1). Scenario 4 used a value of $N_{i,k} = 1.5$ while Scenario 5 used a value of $N_{i,k} = 0.5$.[*] Both scenarios used $K_{i,k} = 1.35 \times 10^{-6}$ as used previously for $Kd_{i,k}$. ($K_{i,k}$ and $N_{i,k}$ are assumed constant in space and time as earlier.) The Euler option with a backward difference and an 0.5-day time step was again used in both scenarios but, because R is now a function of c, the Euler option was executed using the nonlinear, quasi-Newton algorithm (see Chapter 5). Analytical derivatives were used for the Jacobian matrix.

Figure 4.15 shows a vertical snapshot of concentrations at 20 days for both scenarios along with the snapshot assuming linear K_d previously (Scenario 3 from Example 4.2, i.e., $N = 1$) for comparison. Relative to the previous Scenario 3, Scenario 4 ($N = 1.5$) results in a so-called "spreading" front while Scenario 5 ($N = 0.5$) results in a "self-sharpening" front (Fetter, 1999).

Another popular nonlinear isotherm is the Langmuir isotherm, presented in Box 4.2.

BOX 4.2   LANGMUIR ISOTHERM

Another popular nonlinear isotherm is the Langmuir model

$$\bar{c}_{i,k} = \frac{abc_{i,k}}{1 + ac_{i,k}}$$

where a and b are constants. The Langmuir isotherm is put into the GEM's R format as

$$\frac{d(R_{i,k} V_i c_{i,k})}{dt} = (\text{transport} \pm \text{othersources/sinks})$$

where

$$R_{i,k} = 1 + \frac{BD_i ab}{\theta_i (1 + ac_{i,k})}$$

---

[*] As described in Chapters 2 and 5, the Newton algorithm requires a starting "guess" for the candidate concentrations from which its iterations begin. For dynamic problems, all Newton iterations subsequent to the initial time step use the previous solution for this guess. For the initial time step, the candidate concentrations are read from the user-supplied "Guess.csv" file instead of the initial condition file, Initial.csv. This is because it is common to use zero for initial concentrations, yet the value zero can cause problems for Newton's method. For example, for $N = 0.5$ in this problem, a "guess" of zero would result in an illegal function call in evaluating R, i.e,. $0^{-0.5}$ returns a division by zero. The "Guess.csv" approach can avoid this problem by using initial conditions that are zero for all practical purposes, but are not precisely zero. In this GEM run, we used 10^{-8} for the initial guess values to avoid this problem.

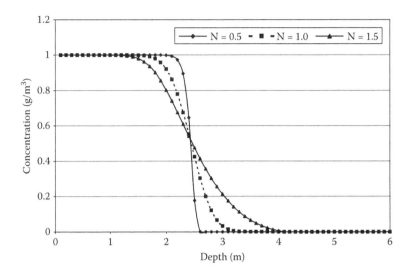

FIGURE 4.15   Results of nonlinear sorption example.

REFERENCES

Fetter, C.W. 1999. *Contaminant Hydrology*, 2nd ed. Upper Saddle River, NJ: Prentice Hall.
Schnoor, J.L. 1996. *Environmental Modeling: Fate and Transport of Pollutants in Water, Air, and Soil*, New York: John Wiley & Sons.

5 Solution Techniques for Steady-State Problems

Recall from Chapter 4 that our generalized system of mass balance differential equations written in matrix form is

$$\frac{d(\mathbf{R}(\underline{c})\mathbf{V}\underline{c})}{dt} = \mathbf{A}\underline{c} + \underline{f}(\underline{c}) + \underline{b} \qquad \text{(4.22 repeated)}$$

For steady-state problems, the chemical mass balance equations comprise a set of simultaneous (coupled), algebraic equations. For a linear problem, the system is

$$\underline{0} = \mathbf{A}\underline{c} + \underline{b}$$

and, for a nonlinear system

$$\underline{0} = \mathbf{A}\underline{c} + \underline{f}(\underline{c}) + \underline{b}$$

Because there is one mass balance equation per chemical per compartment, we are always dealing with the same number of unknowns (concentrations) and number of equations. That is, we always have N equations in N unknowns (N = NEQN in our GEM notation) or a symmetric problem that typically[*] has a unique solution—and the unique solution is what the GEM seeks.

In Section 5.1, we consider these solution techniques for linear problems. Section 5.2 presents methods for nonlinear problems. Finally, Section 5.3 describes and quantifies the characteristics of errors introduced by solving these equations.

In addition to solving systems of algebraic equations for steady-state problems, some of the *dynamic* methods discussed in Chapter 6 also require solving systems of algebraic equations at each time step. The techniques presented in this chapter are also used in those dynamic applications.

5.1 LINEAR SYSTEMS

Solving N linear equations in N unknowns is perhaps the most common task in applied mathematics and is wonderfully simple. Unfortunately, solving such a system with more than two or three equations is a daunting task to many prospective modelers—but only because they have not been exposed to the general procedure called

---

[*] Symmetric problems are typical in applied mathematics, although it is possible to have more unknowns than equations (generally involving an infinite number of solutions) or more equations than unknowns (generally with a single solution).

Gaussian elimination. It was named for Carl Friedrich Gauss, arguably the most influential mathematician in history, who developed the method in the 18th century (it probably took him 10 minutes).

EXAMPLE 5.1

Solving Linear Systems

The best way to show the procedure is by example. Let's consider a simple 3×3 system of equations:

$$2x_1 - x_2 + x_3 = 2 \tag{5.1}$$

$$x_1 + x_2 + 3x_3 = 0 \tag{5.2}$$

$$3x_1 - 2x_2 + 6x_3 = -2 \tag{5.3}$$

where the x variables represent the three unknowns. To solve this system, we want to perform operations on the equations that do not change the relationships among the variables but do put the equations in forms more amenable to solution.

One such operation is multiplication. Everyone knows that multiplying both sides of an equation by the same constant does not change the relationships of the variables. Perhaps less well known is the addition (or subtraction) of two equations. (Adding two equations means adding their respective coefficients and their right-hand-sides.) This operation also does not change the relationships of the variables.* These two operations—multiplication and addition—constitute Gaussian elimination.

Let's first operate on Equation (5.3) with Gaussian elimination so that we eliminate x_1 and x_2, i.e., make their coefficients 0s. Multiplying Equation (5.2) by –3 and adding the resulting equation to (5.3) gives

$$-3x_1 - 3x_2 - 9x_3 = 0 \qquad \text{[–3 times Equation (5.2)]}$$

$$3x_1 - 2x_2 + 6x_3 = -2 \qquad \text{(5.3 repeated)}$$

$$0x_1 - 5x_2 - 3x_3 = -2 \qquad \text{(5.4)}$$

We now replace the original Equation (5.3) with its simpler equivalent (5.4) and our new (and equivalent) system of equations is

$$2x_1 - x_2 + x_3 = 2 \qquad \text{(5.1 repeated)}$$

$$x_1 + x_2 + 3x_3 = 0 \qquad \text{(5.2 repeated)}$$

$$-5x_2 - 3x_3 = -2 \qquad \text{(5.4 repeated)}$$

---

* We leave it to the reader to verify this statement after the problem is solved.

We now give x_1 in (5.2) a coefficient of 0. Multiplying Equation (5.2) by -2 and adding that result to (5.1) gives

$$2x_1 - x_2 + x_3 = 2 \qquad \text{(5.1 repeated)}$$

$$-2x_1 - 2x_2 - 6x_3 = 0 \qquad \text{[-2 times (5.2)]}$$

$$0x_1 - 3x_2 - 5x_3 = 2 \qquad \text{(5.5)}$$

Now replace the original Equation (5.2) with its simpler equivalent (5.5) so our new system of equations is

$$2x_1 - x_2 + x_3 = 2 \qquad \text{(5.1 repeated)}$$

$$-3x_2 - 5x_3 = 2 \qquad \text{(5.5)}$$

$$-5x_2 - 3x_3 = -2 \qquad \text{(5.4 repeated)}$$

Now make the coefficient of x_2 in (5.4) a 0. Multiplying Equation (5.5) by $-5/3$ and adding the result to (5.4) gives

$$5x_2 + \tfrac{25}{3} x_3 = -\tfrac{10}{3} \qquad \text{[-5/3 times (5.5)]}$$

$$-5x_2 - 3x_3 = -2 \qquad \text{(5.4 repeated)}$$

$$\tfrac{16}{3} x_3 = -\tfrac{16}{3} \qquad \text{(5.6)}$$

where we can trivially solve (5.6) for x_3 to get $x_3 = -1$. Our equivalent and much simpler system is now

$$2x_1 - x_2 + x_3 = 2 \qquad \text{(5.1 repeated)}$$

$$-3x_2 - 5x_3 = 2 \qquad \text{(5.5 repeated)}$$

$$x_3 = -1 \qquad \text{(5.6 repeated)}$$

and it is obvious (now that $x_3 = -1$) that we can easily solve (5.5) for x_2 as $-3x_2$ $-5(-1) = 2$ or $x_2 = 1$. Knowing x_3 and x_2, we similarly solve (5.1) to get $x_1 = 2$.

That's it.* Obviously, what we have done is use fundamental multiplication and addition operations on our system of equations to put them into a triangular format

---

* Actually, that's not quite it, but it's good enough for our purposes. Complications may arise when Gaussian elimination results in coefficients of some equation(s) are zero. This is a case of linear dependence among the equations and generally means that the problem is not properly formulated. It is also possible that a problem is properly formulated and the equations are nearly but not quite linearly dependent, and "ill-conditioning" of the matrices poses a problem for solutions.

$$
\begin{bmatrix} 2 & -1 & 1 \\ 1 & 1 & 3 \\ 3 & -2 & 6 \end{bmatrix} \begin{bmatrix} X_1 \\ X_2 \\ X_3 \end{bmatrix} = \begin{bmatrix} 2 \\ 0 \\ -2 \end{bmatrix}
$$

FIGURE 5.1   Example 5.1 equations in matrix form.

$$
\begin{bmatrix} 2 & -1 & 1 \\ 0 & -3 & -5 \\ 0 & 0 & 1 \end{bmatrix} \begin{bmatrix} X_1 \\ X_2 \\ X_3 \end{bmatrix} = \begin{bmatrix} 2 \\ 2 \\ -1 \end{bmatrix}
$$

FIGURE 5.2   Example 5.1 equations in upper triangular form.

so that we can directly solve the last equation for the last unknown, and then work our way back through the equations by back-substitution, solving each equation in turn. If we had 3,000 equations in 3,000 unknowns, this same procedure would lead to their solution. Of course, that would be impossible as a practical matter to do manually as we have here. We would need an algorithm to implement the Gaussian elimination procedure and run it on a computer.

Gaussian elimination* is performed by computer algorithms by putting the equations into matrix form (see Appendix). Our original system of equations in matrix form is shown in Figure 5.1 The Gaussian algorithm then operates on the coefficient matrix and the right-hand side vector (as calculated above by hand) to get them into the upper triangular form as shown in Figure 5.2, and then obtains the solution by back substitution.

The linear equation solver in the GEM uses Gaussian elimination to factor the A matrix into triangular form and then solves for the concentrations by back substitution. It operates slightly differently from our simple example above in that both upper triangular and lower triangular decomposition results, but for our purposes here it is similar in principle to the above example. The equation solver consists of two subroutines—LUDCMP and LUBKSB—both taken from the gold standard of scientific computing, *Numerical Recipes* (Press et al., 1986). LUDCMP performs the lower and upper decomposition while LUBKSB performs the back substitution.

For readers who acquire the GEM source code, these two sub-routines will not be included because their source code is copyrighted. The BASIC versions of these subroutines are (still) available from Cambridge University Press (http://www.cambridge.org/us/catalogue/catalogue.asp?isbn=0521406897) and the cost was $75

---

* The Gaussian elimination method shown here is a direct method, i.e., once the equations are transformed into an alternative structure, the solution is then available. There are other methods for solving linear systems based on iterative techniques, somewhat analogous to our upcoming discussion of Newton's method for nonlinear equations. These methods, including the Jacobi, Gauss-Seidel, and successive over-relaxation (SOR) methods are sometimes computationally advantageous for very large systems of equations.

as of this writing. However, the *Numerical Recipes* sub-routines have worked well for us over the years and we highly recommend them. (We also recommend the addition of *Numerical Recipes* (Press et al., 1986) to the library of every prospective modeler.) For readers who acquire the GEM source code, we provide an alternative linear solver that is open-source. It was developed by Jon Squire at the University of Maryland Baltimore County (UMBC). Squire's algorithm uses Gaussian elimination to extend the conversion of the matrix system beyond the upper-triangular format to the point where the $\mathbf{A}$ matrix becomes the identity matrix, $\mathbf{I}$, and the solution can be read directly from that form, i.e., without backward substitution. Comparisons of this algorithm's results with those using the *Numerical Recipes* sub-routines show identical results to many significant figures. The *Numerical Recipes* approach, however, is more computationally efficient, requiring fewer mathematical operations.

5.2   NONLINEAR SYSTEMS

Unlike the linear case, solving systems of nonlinear algebraic equations is not marvelously simple. Iterative methods and initial guesses are needed and numerical and convergence issues may be problematic. Nonetheless, the theory behind the most popular method—the Newton-Raphson[*] method—is remarkably succinct, easily understandable, and requires only knowledge of a derivative. When the method is applied to a single equation, matrix notation can be avoided. However, for our purposes for multiple equations, we again need to use matrix notation.

5.2.1   CLASSICAL NEWTON'S METHOD

As described in many references (e.g., Matthews, 1992), the classical Newton-Raphson algorithm (sometimes simply called Newton's method) for solving systems of nonlinear algebraic equations begins by expressing the system of equations in implicit form, i.e., expressing all equations as equal to 0:

$$\underline{\mathbf{g}}(\underline{\mathbf{c}}) = \underline{\mathbf{0}} \tag{5.7}$$

where $\underline{\mathbf{g}}$ is an NEQN × 1 vector function of $\underline{\mathbf{c}}$, the set of NEQN variables (concentrations) to be solved. After a nonlinear system of equations is expressed in implicit form as above, finding $\underline{\mathbf{c}}$ is equivalent to finding the roots of the system of equations.

---

[*] Continuing our tour of ancient mathematicians (most of what we need to know was discovered very long ago!). According to www.Wikipedia.com, Joseph Raphson "was an English mathematician known best for the Newton-Raphson method. Little is known about his life, and even his exact years of birth and death are unknown, although the mathematical historian Florian Cajori provided the approximate dates 1648–1715. Raphson's most notable work is *Analysis Aequationum Universalis*, published in 1690. It contains a method, now known as the Newton-Raphson method, for approximating the roots of an equation. Isaac Newton had developed a very similar formula in his *Method of Fluxions*, written in 1671, but this work would not be published until 1736, nearly 50 years after Raphson's *Analysis*. However, Raphson's version of the method is simpler than Newton's, and is therefore generally considered superior. For this reason, it is Raphson's version of the method, rather than Newton's, that is to be found in textbooks today."

Beginning with an initial estimate of $\underline{c}_0$ for $\underline{c}$, the Newton-Raphson algorithm proceeds as described in the following for iteration k.

> *Step 1.* Evaluate the function for $\underline{c}_k$, i.e., $\underline{g}(\underline{c}_k)$.
> *Step 2.* Is $\underline{g}(\underline{c}_k)$ approximately equal to 0 within some convergence criterion? If yes, then stop; the solution has been obtained. If no, go to Step 3.
> *Step 3.* Update the solution vector as

$$\underline{c}_{k+1} = \underline{c}_k + \underline{\Delta c}_k \tag{5.8}$$

where $\underline{\Delta c}_k$ is the solution to the linear system of equations (Newton step)

$$\mathbf{J}(\underline{c}_k)\underline{\Delta c}_k = -\underline{g}(\underline{c}_k) \tag{5.9}$$

and $\mathbf{J}$ is the NEQN × NEQN Jacobian matrix of partial derivatives, i.e.,

$$J_{ij} = \frac{\partial g_i}{\partial c_j} \tag{5.10}$$

If analytical derivatives are not available or easily obtainable, the Jacobian elements can be approximated by numerical finite difference methods. The GEM includes this option using a central differencing method, wherein the user specifies the finite difference parameter to use. (Approximating the partial derivatives numerically requires *substantially* more computation than using analytical derivatives. For relatively small problems, the user will not notice the difference. However, for larger problems (even NEQN = 100) this step can be a *major* time hog. Analytical derivatives should be used whenever possible and practicable.) Equation (5.7), written in terms of the GEM steady-state model, is

$$\underline{g}(\underline{c}) = \mathbf{A}\underline{c} + \underline{f}(\underline{c}) + \underline{b} = \underline{0} \tag{5.11}$$

and the *i,j*th element of the Jacobian matrix is

$$\frac{\partial g_i}{\partial c_j} = A_{i,j} + \frac{\partial f_i}{\partial c_j} \tag{5.12}$$

where $A_{i,j}$ denotes the *i,j*th element of the $\mathbf{A}$ matrix and $\partial f_i / \partial c_j$ is the partial derivative of the *i*th row of the nonlinear source or sink vector $\underline{f}(\underline{c})$ with respect to c_j. If a finite difference approximation to the Jacobian matrix is specified, all partial derivatives [Equation (5.12) for all *i* and *j*] are estimated numerically by the GEM. If analytical derivatives are preferred, the user must supply the analytical derivatives $\partial f_i / \partial c_j$ for all *i* and *j*. Box 5.1 provides a graphical interpretation of Newton's method for solving one equation in one unknown.

BOX 5.1 GRAPHICAL INTERPRETATION OF NEWTON'S METHOD

Consider finding the root of a single equation in one unknown as depicted in Figure 5.3. The root is that value of c, c^* where the g function crosses the c axis, i.e., where $g(c^*) = 0$. A single iteration of the Newton algorithm is illustrated in the figure, starting with the value of c at iteration k, c_k. At the end of iteration k, the updated value of c is c_{k+1}, and indeed it is closer to c^* than c_k. The heart of the Newton algorithm is to construct a tangent line to the g function at c_k, follow that tangent down (or up, depending on where we start) until it crosses the c axis. That point then becomes the updated c_{k+1}. The process is repeated at c_{k+1}, c_{k+2}, etc. until the g function is approximately 0 (within some convergence criterion). The Newton algorithm can be constructed by considering the equation for the slope of the tangent line:

$$m = \frac{\Delta g}{\Delta c} = \frac{0 - g(c_k)}{c_{k+1} - c_k}$$

Solving for $c_k + 1$

$$c_{k+1} = c_k - \frac{g(c_k)}{m}$$

where

$$m = \frac{dg(c_k)}{dc},$$

i.e., the derivative of the g-function at c_k.

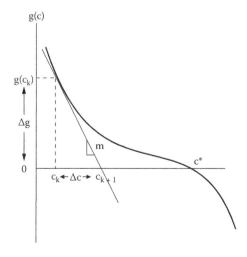

FIGURE 5.3 Graphical interpretation of Newton's method.

5.2.2 QUASI-NEWTON METHOD AND RANDOM START ENHANCEMENTS TO NEWTON'S METHOD

Solving nonlinear systems of equations is often a fickle process, and using Newton's method is no exception. When it works, it works extremely well (rapid convergence) relative to alternatives such as optimization methods, but its success is highly dependent on the starting point for the iterative algorithm[*] (initial guess from which the iterations begin).

For steady-state problems in the GEM, the user must provide the initial guess, because the Newton algorithm simply has no information from which to estimate a reasonable starting point. (The guess doesn't *necessarily* have to be close to the solution, but the closer the better.) For dynamic problems, as Newton's method is used to advance the solution from time t to t + 1, the previously determined solution at t is obviously a good choice as a starting point to determine the solution at t + 1 and is used as the guess. There is obviously a trade-off between the length of the time step and the closeness of the previous solution to the new solution. While large time steps may be desirable computationally for some problems, they may also result in non-convergence. However, the user is prompted for the initial guess[†] for the first time step, i.e., the guess to advance the solution from the initial condition to the next time step.

The advantage of the classical Newton method described in Section 5.2.1 is that the convergence attained is very fast (quadratic convergence). The disadvantage is that the algorithm is local, i.e., in general, the starting point must be in relatively close proximity to the solution; otherwise the algorithm may simply bog down (go nowhere) or blow up. Quasi-Newton methods modify the Newton step (see Section 5.2.1) to ensure that the step taken is not so large as to make the next solution update in the iterative algorithm "worse" than the current solution update. Recall from Equation (5.8) that the updated solution vector is calculated as

$$\underline{c}_{k+1} = \underline{c}_k + \underline{\Delta c}_k$$

where $\underline{\Delta c}_k$ is the Newton step. How would we define a "worse" update? A reasonable approach is to consider that finding the solution to the system of nonlinear equations

$$\underline{g}(\underline{c}) = \mathbf{0}$$

---

[*] Indeed, Press et al. (1986) go so far as to say: "There are *no* good, general methods for solving systems of more than one nonlinear equation. Furthermore ... there *never will be* any good, general methods." [They then demonstrate graphically why.] "For problems in more than two dimensions ... root finding becomes virtually impossible without insight" [regarding the starting guess].

[†] Although the already user-supplied initial condition may seem analogous to using the solution at t as the guess for t + 1 for the other time steps, initial conditions of zero (which are common) can cause problems for the algorithm, e.g., it may never "escape" from zero or, for solutions using finite difference estimates of the Jacobian, division by zero, an error will occur. Therefore, a non-zero, user-specified guess must be entered.

is equivalent to finding the minimum of a general quadratic function f^*

$$f(\underline{c}) = \frac{1}{2}\sum_i g_i(\underline{c})^2 = \frac{1}{2}\underline{g}(\underline{c})^T\underline{g}(\underline{c}) \tag{5.13}$$

where T denotes the transpose operator, i.e., transposing columns to rows or vice-versa.

These are equivalent problems because the necessary condition for a minimum of the quadratic function f is that all the elements of its gradient (vector of first partial derivatives) must be zero at its minimum, that is

$$\underline{\nabla f}(\underline{c}^*) = \left[\frac{\partial f}{\partial c_1}\Big|_*, \frac{\partial f}{\partial c_2}\Big|_*, \dots, \frac{\partial f}{\partial c_{NEQN}}\Big|_* \right]^T = \underline{0}^T \tag{5.14}$$

where c^* denotes a concentration that minimizes f. For example, the partial derivative of f with respect to an arbitrary jth concentration is, from (5.14)

$$\frac{\partial f(\underline{c})}{\partial c_j} = \sum_i^{NEQN}\left(g_i(\underline{c})\frac{\partial g_i}{\partial c_j} \right) \tag{5.15}$$

so that, if all the g_i terms are zero (thus satisfying our original problem above), all the partial derivatives will also be zero. Given this equivalency,[†] a natural criterion for evaluating the usefulness of an update under consideration, $\underline{c}_{k+1}$, relative to the current $\underline{c}_k$ is

$$f(\underline{c}_{k+1}) < f(\underline{c}_k) \tag{5.16}$$

so that we are in fact decreasing f by this update and on our way to minimizing it. Accordingly, we introduce a scalar[‡] parameter λ into the Newton update equation so that the update is now

$$\underline{c}_{k+1} = \underline{c}_k + \lambda\underline{I}\underline{\Delta c}_k \tag{5.17}$$

---

[*] The f function in this section denotes an arbitrary quadratic function. It is *not* the $f(\underline{c})$ nonlinear source/sink function used in the GEM, i.e., Equation (5.11).

[†] The reader may wonder why we don't simply solve the *minimization* problem using Newton's method, i.e., solve $\underline{\nabla f}(\underline{c}) = \underline{0}$ and use the matrix of second partial derivatives of f (the Hessian matrix) in lieu of the Jacobian matrix to estimate our Newton step. This could be done, but Dennis and Schnabel (1983) advise that "… it is better to use the structure of the original problem whenever possible, in particular to compute the Newton step." While it is true that every solution to the system of equations problem (there may be multiple roots) is also a solution to the corresponding minimization problem, the converse is not true. (There may be local minima that do not correspond to roots.) In addition, use of the system's Jacobian to find the Newton step is preferable to use of the minimization problem's Hessian.

[‡] A scalar is a single-valued variable, i.e., not a vector or a matrix.

where I is the NEQN × NEQN identity matrix. We are simply multiplying each element of the Newton step vector by the scalar λ. Because non-convergence problems may arise from the Newton step taking us too far, we require

$$0 < \lambda \leq 1.0 \qquad (5.18)$$

Therefore, the quasi-Newton algorithm works at each iteration by first calculating the Newton step as described previously (Section 5.2.1). However, before making the full Newton step (i.e., $\lambda = 1$), we apply the criterion (5.16) to see whether an improvement results. If so, we take the full Newton step.* If not, we must then decide on some value for λ that will result in improvement.

An appropriate value of λ can be determined in a variety of ways (Strang, 1986; Dennis and Schnabel, 1983). We adopted a particularly simple method using bisection. That is, if $\lambda = 1$ is not an improvement, we set $\lambda = \frac{1}{2}$. If $\lambda = \frac{1}{2}$ is not an improvement, we set $\lambda = \frac{1}{4}$, and so on. A major motivation for this approach is another, more pragmatic issue that can arise with the updates: they can become negative. Negative concentrations are nonsensical in the context of the GEM (although we would be perfectly happy to use them temporarily as a means to a speedy convergence to a true solution) and can cause illegal mathematical operations, e.g., exponentiation of a negative number—a real "show stopper" in computing. For problems that present this issue, we had success with the bisection method. Therefore, it seems prudent to use the same bisection method to address both the negative concentration problem and the need to ensure progress in minimizing f before making a step.

With the introduction of the quasi-Newton modification, the GEM's nonlinear capabilities are greatly enhanced with respect to robustness of initial guesses. Nonetheless, two additional problems can arise.† First, during the bisection algorithm to determine an f-decreasing λ value, the bisection can ultimately result in a value that is sufficiently small as to not be computationally distinguishable from zero. Clearly, a λ of zero will not advance the algorithm.

The actual value at which this occurs is machine-specific and is called the machine epsilon. Should the bisection reach the machine epsilon, something must be done. (Determining the machine-specific epsilon is performed internally and automatically within the GEM. For our current computer, that value is approximately 10^{-17}.)

The second problem is that some starting points (initial guesses) for some systems of equations are simply poor places to begin, not for the reason that a decrease in f cannot be made from those points (i.e., the bisection will find some λ above the machine epsilon that is a decrease), but the decreases can be excruciatingly slow.

Should either of the problems occur during the quasi-Newton iteration, the GEM invokes a random sampling method from which a new guess is generated and the iteration is restarted. From N random samples of the values of the solution vector

---

* It is extremely important to first attempt $\lambda = 1$. This ensures that the quadratic convergence advantage of classical Newton remains when we are in the vicinity of a solution and full Newton steps are possible.

† A potential third problem is that non-convergence can arise from an ill-conditioned or non-positive definite Jacobian matrix, usually resulting from a poorly specified problem. There are numerical methods to anticipate and correct for this condition, but they are not included in this version of the GEM.

to be updated, the GEM determines which of the N guesses has the lowest f-value (Equation 5.13). This $\underline{c}$ is then used as the guess and the algorithm is restarted. The user specifies the number of random samples to be drawn and the concentration range within which the (uniform) sampling is performed. The user also specifies the number of consecutive times that the random sampling fall-back procedure can be invoked before the GEM admits failure and errors out (in which case, the user should use a smaller step size and try again).

EXAMPLE 5.2

Classical Newton's Method versus Quasi-Newton Method

The GEM's quasi-Newton and random sampling enhancements are demonstrated in two examples taken from the literature. These are non-environmental modeling examples and were solved using the *Equation Solver* mode. The first example is solving a system of nonlinear equations while the second solves a minimization problem, which is also equivalent to solving a system of nonlinear equations as previously described.

The first example is taken from Dennis and Schnabel (1983) and demonstrates the failure of the "local" classical Newton method versus the globally convergent quasi-Newton modification. The problem is to find the solution to the system of two equations in two unknowns

$$g_1(\underline{x}) = x_1^2 + x_2^2 - 2 = 0$$

$$g_2(\underline{x}) = e^{x_1 - 1} + x_2^3 - 2 = 0$$

starting from the point (initial guess) $x_1 = 2$ and $x_2 = 0.5$. The known solution is $x_1 = 1$ and $x_2 = 1$. We set this problem up in the GEM (using an analytical Jacobian) and first ran it using the classical Newton algorithm, i.e., λ fixed at 1.0 from the above initial guess. With this classical Newton approach, the algorithm diverged (increasing f) immediately from the starting point, ultimately resulting in an overflow error after 16 Newton iterations.

We ran the problem again from the same starting point, this time allowing λ to decrease in accordance with the quasi-Newton algorithm. We specified a stopping criterion of $f \leq 10^{-10}$. For the first two Newton iterations, λ was decreased by the algorithm from the initial 1.0 to 0.0078125 and 0.125, respectively. After that, the solution was apparently in close enough proximity to the true roots that a full Newton step could be made and λ remained at 1.0 for all subsequent iterations. Convergence was attained after seven iterations. Thus, the quasi-Newton approach clearly allowed this problem to be solved and the classical Newton approach did not.

The second problem is a classical minimization reported throughout the operations research literature. It involves trying to find the minimum of the "banana function." The banana function was pathologically constructed (Rosenbrock, 1960) to result in contours of equal f-value that resemble a banana, but more usefully described as a long, deep, narrow canyon. If one is using a descent method (the Newton or quasi-Newton) from a starting point that lies at an "entrance" to this canyon, the descent iterations result in successive decreases, but only very marginally so.

The narrowness of the canyon causes the successive iterations to "bounce off" opposing canyon walls without much downward progress. The banana function is used to demonstrate very slow convergence of descent-based optimization methods for such problems given poor starting points. (Note that the GEM is not set up to solve optimization problems per se, but we can easily solve them by calculating analytically the gradient of the objective function and treat that as a system of equations to be solved. The Jacobian can be estimated by finite differences or the Hessian matrix [matrix of second partial derivatives of the objective function] can be substituted for the GEM's Jacobian.)

The f function constituting the banana function is $f = 100(x_1^2 - x_2)^2 + (1 - x_1)^2$ with a known minimum $f = 0$ at (1,1). The gradient we used as our system of equations for which to find the roots is

$$\frac{\partial f}{\partial x_1} = 400x_1^3 - 400x_2x_1 + 2x_1 - 2 = 0$$

$$\frac{\partial f}{\partial x_2} = -200(x_1^2 - x_2) = 0$$

The (analytical) Hessian matrix of second partial derivatives (that we substituted for the GEM's Jacobian matrix) is

$$H_{1,1} = \frac{\partial^2 f}{\partial x_1^2} = 1200x_1^2 - 400x_2 + 2$$

$$H_{1,2} = \frac{\partial^2 f}{\partial x_1 \partial x_2} = H_{2,1} = \frac{\partial^2 f}{\partial x_2 \partial x_1} = -400x_1$$

$$H_{2,2} = \frac{\partial^2 f}{\partial x_2^2} = 200$$

The banana function problem was first run using the GEM with the quasi-Newton algorithm implemented but without the random sampling enhancement. The starting points were $x_1 = 6.39$ and $x_2 = -0.221$, which are popular starting points at an entrance to the canyon (Dennis and Schnabel, 1983). The f-value associated with the starting point is 5.5×10^9. The first iteration made a full Newton step to achieve a significant decrease resulting in $f = 58.1$ and updated $x_2 = 40.8$. (x_1 was not significantly altered.) After that first iteration, however, another 500 iterations served only to decrease the f-value to 56.4. All the subsequent iterations also involved decreases in λ. Clearly, even the quasi-Newton method is agonizingly slow at solving this problem given the poor starting point.

The second GEM run of the banana function allowed the random sampling procedure to be invoked. One hundred random samples were drawn between −5 and 5 (for both x values). After the quasi-Newton algorithm ran its course (again showing very little improvement), the random sampling algorithm drew a new starting point of $x_1 = 1.23$ and $x_2 = 1.48$ with an associated f-value of 110.4 (representing the best of the 100 random samples). Although this starting point had a much greater f-value than the ending point of the quasi-Newton phase, it is apparently much better poised to effect a rapid decrease to the minimum of the banana function. From this starting point, 18 additional quasi-Newton iterations were needed to satisfy a stopping criterion of $f \le 10^{-10}$. Thus, the random sampling enhancement to the quasi-Newton method to select new starting points adds much additional flexibility to the GEM.

5.3   ACCURACY OF SOLUTIONS

By way of review, the compartment modeling approach is typically undertaken as a numerical means of solving partial differential equations that do not have analytical solutions. These underlying differential equations are the true mathematical models and, if analytical solutions were available, those solutions would be considered completely accurate, at least to the extent that the differential equations completely represent the physical prototype being simulated.

The compartment modeling approach discretizes the true equations (actually it discretizes the physical system) and the discretization introduces errors. These errors are discussed in this section for steady-state linear conditions and are associated with the choice of the spatial differencing parameter α. Temporal discretization for dynamic problems introduces additional errors and issues that are discussed in Chapter 6. Error characteristics for nonlinear problems are beyond our scope; they must be analyzed "locally" using approximate methods.

If your system has "real" completely mixed compartments and the GEM compartment modeling approach simply reflects the physical system, discretization errors are not issues and the accuracy issues discussed in this section may not be applicable.[*]

5.3.1   DISCRETIZATION ERROR AND NUMERICAL DISPERSION

The most general (three-dimensional, varying coefficients) form of the underlying, steady-state, linear differential equation that the GEM approximates is

$$0 = -\frac{1}{A_x}\frac{d}{dx}(Q_x c) - \frac{1}{A_y}\frac{d}{dy}(Q_y c) - \frac{1}{A_z}\frac{d}{dz}(Q_z c) + \frac{1}{A_x}\frac{d}{dx}\left(E_x A_x \frac{dc}{dx}\right) +$$
$$\frac{1}{A_y}\frac{d}{dy}\left(E_y A_y \frac{dc}{dy}\right) + \frac{1}{A_z}\frac{d}{dz}\left(E_z A_z \frac{dc}{dz}\right) \pm kc \tag{5.19}$$

For simplicity, however, we assume here that the true underlying model is the one-dimensional version of (5.19) with spatially constant parameters given by

$$0 = -\overline{U}\frac{dc}{dx} + E\frac{d^2 c}{dx^2} \pm kc \tag{5.20}$$

where we introduce for simplicity $\overline{U}$ as the flow velocity U (L/T) normalized by the water content θ, i.e., $\overline{U} = U/\theta$, as appropriate for porous media systems.

---

[*] Or they may be applicable. For example, if you use a central spatial difference, you would have to worry about positivity or wiggle whether your discretization is real or not.

For use in subsequent examples, for a loading (W) at $x = 0$, where positive x is downstream of the loading, negative x is upstream, and boundary conditions of 0 are plus and minus infinity, Equation (5.20) has the exact solution

$$c(x) = \begin{array}{l} c(0)e^{j_1 x}, \text{ for } x \le 0 \\ c(0)e^{j_2 x}, \text{ for } x \ge 0 \end{array} \tag{5.21a}$$

where

$$c(0) = \frac{W}{Q\beta} \tag{5.21b}$$

$$j_1 = \frac{\overline{U}}{2E}(1+\beta) \tag{5.21c}$$

$$j_2 = \frac{\overline{U}}{2E}(1-\beta) \tag{5.21d}$$

and

$$\beta = \sqrt{1 + \frac{4kE}{\overline{U}^2}} \tag{5.21e}$$

It is easier to understand the discretization errors if we work with a finite difference approximation to Equation (5.20) *itself*, rather than discretizing the spatial domain. For constant parameters, the two approaches (finite difference versus compartments) are equivalent as first demonstrated by Thomann (1972).

A finite difference approximation to Equation (5.20) replaces the derivative terms by approximations to those derivatives. For example, a backward difference approximation to the first derivative is

$$\frac{dc}{dx} \approx \frac{c(x) - c(x - \Delta x)}{\Delta x} \tag{5.22}$$

where Δx is the finite difference. Suppose we replaced the term

$$-\overline{U}\frac{dc}{dx}$$

with the backward difference term

$$-\overline{U}\frac{c(x) - c(x - \Delta x)}{\Delta x}.$$

What discretization error have we introduced?

We can approximately quantify this error using a Taylor series approach (see Box 5.2). Let us expand $c(x - \Delta x)$ around $c(x)$ in a Taylor series, truncated after the second derivative term.

$$c(x - \Delta x) = c(x) - \Delta x \frac{dc}{dx} + \frac{\Delta x^2}{2} \frac{d^2c}{dx^2} + \vartheta(\Delta x^3) \qquad (5.23)$$

where $\vartheta(\Delta x^3)$ denotes the relative magnitude or order of the error introduced by the truncation. That is, the error will be on the order of Δx^3. Our differential equation with a backward difference approximation to the first derivative term is

$$0 = -\bar{U} \frac{c(x) - c(x - \Delta x)}{\Delta x} + E \frac{d^2c}{dx^2} \pm kc \qquad (5.24)$$

Substituting from (5.23) for $c(x - \Delta x)$ in Equation (5.24) we have

$$0 = -\bar{U}\left(\frac{c(x) - \left(c(x) - \Delta x \frac{dc}{dx} - \frac{\Delta x}{2} \frac{d^2c}{dx^2} \right)}{\Delta x} \right) + E \frac{d^2c}{dx^2} \pm kc + \vartheta(\Delta x^3) \qquad (5.25)$$

BOX 5.2   TAYLOR SERIES

The Taylor series is probably the most important tool in the field of numerical methods. The series allows the value of a function at some point x to be exactly estimated by an infinite power series applied at a nearby point a. While a reader may legitimately question the utility of trying to build an *infinite* power series, the real beauty of the Taylor series is that the value of the function can generally be approximated to a high degree of accuracy with only a few terms of the Taylor series if points x and a are reasonably close. Moreover, the series allows the truncation error to be approximately quantified. While the use of a Taylor series to approximate the value of a function is useful in many applications, for our purposes the most important role of the Taylor series is in approximating the derivatives of a function and be able to do so within an approximately known error level.

The Taylor series expansion of a function of x, $f(x)$, around a nearby point a is:

$$f(x) = f(a) + (x - a)\frac{df}{dx} + \frac{(x-a)}{2!}\frac{d^2f}{dx^2} + \frac{(x-a)^3}{3!}\frac{d^3f}{dx^2} + ... + \frac{(x-a)^n}{n!}\frac{d^nf}{dx^n} + ...$$

If a Taylor series is truncated, the error introduced by the omitted terms is approximately known. If the series is truncated after the term containing $(x - a)^n$, the error is known to be "on the order of" $(x - a)^{n+1}$ (Hornbeck, 1975).

Consider the function $f(x) = x^3$. Suppose we wish to estimate the value of $f(10)$, which is 1,000, by expanding $f(x)$ in a Taylor series about the point $x = 5$. We can obtain the exact value of $f(10)$ by using the complete Taylor series. The derivatives of $f(x) = x^3$ are

$$\frac{df}{dx} = 3x^2$$

$$\frac{d^2 f}{dx^2} = 6x$$

and

$$\frac{d^3 f}{dx^3} = 6$$

after which all subsequent derivatives are zero. Because the derivatives of $f(x)$ are constant after the third derivative, the complete Taylor series will have four terms:

$$f(10) = 5^3 + (10 - 5)(3)(5^2) + \frac{(10 - 5)^2}{2!}(6)(5) + \frac{(10 - 5)^3}{3!}(6) = 1000$$

which is exact. Suppose, however, we used only the first three terms of the above Taylor series to estimate $f(10)$, i.e.,

$$f(10) \approx 5^3 + (10 - 5)(3)(5^2) + \frac{(10 - 5)^2}{2}(6)(5) = 875$$

We know that the error is "on the order of" $(x - a)^{n+1} = 5^3 = 125$, which in this case is the exact error.

By simplifying and collecting terms, we can write this as

$$0 = -\overline{U}\frac{dc}{dx} + \left(E + \frac{\overline{U}\Delta x}{2}\right)\frac{d^2 c}{dx^2} \pm kc + \vartheta(\Delta x^3) \tag{5.26}$$

If we compare (5.26) to our original differential Equation (5.20), in addition to the truncation error term, we see that the backward finite difference approximation to the first derivative effectively introduced a new term

$$\frac{\overline{U}\Delta x}{2}$$

as a coefficient of the second derivative. This is called numerical dispersion[*] and has the effect of increasing the true dispersion E by the amount

$$\frac{\overline{U}\Delta x}{2}.$$

Suppose we made a forward difference approximation to the first derivative, i.e., our differential equation is now approximated as

$$0 = -\overline{U}\frac{c(x+\Delta x)-c(x)}{\Delta x} + E\frac{d^2c}{dx^2} \pm kc \qquad (5.27)$$

Let us expand c(x+Δx) around c(x) in a Taylor series, truncated after the second derivative term.

$$c(x+\Delta x) = c(x) + \Delta x\frac{dc}{dx} + \frac{\Delta x^2}{2}\frac{d^2c}{dx^2} + \vartheta(\Delta x^{3)} \qquad (5.28)$$

Similar to the backward difference analysis, substituting from (5.28) into (5.27), simplifying and collecting terms, we have

$$0 = -\overline{U}\frac{dc}{dx} + \left(E - \frac{\overline{U}\Delta x}{2}\right)\frac{d^2c}{dx^2} \pm kc + \vartheta(\Delta x^3) \qquad (5.29)$$

and we see that the forward difference approximation also introduces a numerical dispersion term of the same value as the backward difference, but with a negative sign,

$$-\frac{\overline{U}\Delta x}{2}.$$

Thus, a backward difference increases the true dispersion by this amount; a forward difference decreases it by this amount.

For systems that are largely dispersive, i.e., mass transport due to advection is small relative to dispersive mass transport, this numerical dispersion will be small and is not a major concern. However, for highly advective systems, this artificial smearing of the concentrations in the space domain can be a significant error. One method of controlling the solution for this numerical dispersion for a backward difference is to subtract from the true dispersion coefficient E the amount of the numerical dispersion

---

[*] Hoffman (1992) provides a more detailed examination of the causes of numerical dispersion and distinguishes between numerical dispersion and numerical diffusion. We are content here to lump them both into the dispersion term.

$$\frac{\overline{U}\Delta x}{2}$$

and use a modified dispersion coefficient

$$E - \frac{\overline{U}\Delta x}{2}$$

in model simulations. For a forward difference, we would add the numerical dispersion.

We introduced the forward spatial difference only to demonstrate completeness. This method has nothing to recommend it by way of any advantage over the backward or central difference method (discussed below). It introduces the same quantity of numerical dispersion as does the backward difference, albeit in a different direction. Therefore, it has no advantage over the backward difference. As will be shown in an upcoming section on positivity and wiggle considerations, it, like the central difference method, suffers from positivity and wiggle constraints. In fact, it is more susceptible to positivity and wiggle than is the central difference. Because it presents no advantage over backward or central differencing schemes, we will not further consider the forward spatial difference.

Another method of eliminating numerical dispersion is to use a central difference approximation to the first derivative instead of either the backward (or forward) difference. A central difference approximation[*] is

$$\frac{dc}{dx} \approx \frac{c(x+\Delta x) - c(x-\Delta x)}{2\Delta x} \tag{5.30}$$

so that our approximated differential equation is now

$$0 = -\overline{U}\frac{c(x+\Delta x) - c(x-\Delta x)}{2\Delta x} + E\frac{d^2c}{dx^2} \pm kc \tag{5.31}$$

and can be shown to have a truncation error on the order of Δx^3. Substituting for $c(x + \Delta x)$ from (5.28) and $c(x - \Delta x)$ from (5.23), we have

$$0 = -\overline{U}\left(\frac{\left(c(x)+\Delta x\frac{dc}{dx}+\frac{\Delta x}{2}\frac{d^2c}{dx^2}\right)-\left(c(x)-\Delta x\frac{dc}{dx}+\frac{\Delta x}{2}\frac{d^2c}{dx^2}\right)}{2\Delta x}\right)$$
$$+E\frac{d^2c}{dx^2} \pm kc + \vartheta(\Delta x^3) \tag{5.32}$$

---

[*] Relative to the GEM's compartment approach and the spatial discretization parameter α, the backward difference corresponds to α = 1, the forward difference is α = 0, and the central difference corresponds to α = 0.5 for equal size compartments.

We see this time that the second derivative terms cancel each other and (5.32) simplifies to

$$0 = -\overline{U}\,\frac{c(x+\Delta x)-c(x-\Delta x)}{2\Delta x} + E\,\frac{d^2c}{dx^2} \pm kc + \vartheta(\Delta x^3) \tag{5.33}$$

Note that the dispersion term does not now include a numerical dispersion component. However, while the use of a central difference avoids numerical dispersion, it is not a panacea to accuracy as it can give rise to negative or wiggly concentrations as discussed later.

Note that the use of a central difference approximation to the first derivative is inherently more accurate than a backward or forward difference (Hoffman, 1992). For example, when we substituted the backward difference approximation into the underlying differential equation to result in Equation (5.26) above, we retained the second-order term

$$\frac{\Delta x}{2}\,\frac{d^2c}{dx^2}$$

to show how it adds to the true dispersion; the overall order of the truncation error was shown on the order of Δx^3. When using a backward difference in an actual finite difference approximation to the differential equation, however, this second-order term would be ignored and would add to the discretization error, increasing the error to the order of Δx^2. In contrast, use of the central difference involves an error on the order of Δx^3.

We now make a finite difference approximation to the second-order derivative in Equation (5.20) to complete the numerical solution. We already have the two required Taylor series expansions [(5.23) and (5.28)]. Adding them and noting that all odd-powered terms will cancel, we have

$$c(x+\Delta x)+c(x-\Delta x)=2c(x)+\Delta x^2\,\frac{d^2c}{dx^2}+\vartheta(\Delta x^3) \tag{5.34}$$

which can be solved for the second derivative as

$$\frac{d^2c}{dx^2} = \frac{c(x+\Delta x)-2c(x)+c(x-\Delta x)}{\Delta x^2} + \vartheta(\Delta x^3) \tag{5.35}$$

We can now substitute the finite difference approximations of the two derivatives into the underlying differential Equation (5.20) to see the complete numerical model. We will use the backward difference option for the first derivative. We have now

$$0 = -\overline{U}\,\frac{c(x)-c(x-\Delta x)}{\Delta x} + E\,\frac{c(x+\Delta x)-2c(x)+c(x-\Delta x)}{\Delta x^2} \pm kc(x) \tag{5.36}$$

If we denote c(x) as c_i, $c(x + \Delta x)$ as c_{i+1}, and $c(x - \Delta x)$ as c_{i-1}, analogous to the compartment model nomenclature, multiply through by compartment volume $V = A\Delta x$, and collect terms, we have

$$0 = c_i(-\overline{Q} - E' \pm kV) + c_{i-1}(\overline{Q} + E') + c_{i+1}(E') \tag{5.37}$$

where $\overline{Q}$ is the water content-normalized flow parameter. Equation (5.37) can be seen to be analogous to the form of the GEM's compartment model, i.e., Equation (4.19) at steady state, constant parameters, and $\alpha = 1$, thus showing the equivalence of the finite difference and compartment approaches. This exercise was intended to demonstrate how truncation errors arise as a result of finite difference (or compartment) approximations to the spatial derivatives and how they cause an artificial numerical dispersion for the backward difference option to advective transport.

EXAMPLE 5.3

Illustration of Discretization Errors and Numerical Dispersion

Let us now look at some examples to illustrate discretization error and numerical dispersion by comparing GEM simulations with an analytical solution. The examples come from Thomann and Mueller (1987) and represent the steady-state spatial distribution of a pollutant in a one-dimensional estuary subject to advection, dispersion, and first-order decay. The water content is 1.0 and there is no retardation. The pollutant is introduced at point $x = 0$. Dispersive mass transport carries the pollutant upstream of the loading ($x < 0$), while advection and dispersion affect the concentration downstream ($x > 0$). The analytical solution is given by Equations (5.21) above. The estuary has a uniform flow of 122,328 m³/day and a uniform width and depth of 93 m by 1 m, respectively. The corresponding flow velocity is 1,320 m/day. The uniform dispersion coefficient (E) is 1.3×10^7 m²/day. The loading at $x = 0$ is 1,814,369 g/day. (These oddly precise units in our example result from converting the Thomann and Mueller example in English units to metric units.)

First, it is worth repeating that because discretization errors arise from finite difference approximations to the spatial derivatives, it follows that if there are no concentration gradients, there will be no errors in these derivative approximations and thus no discretization errors (and generally no need to build a numerical model of such a simple steady state system because the solution should be known simply from boundary conditions, loads, and flows).

To illustrate this, we first ran the analytical solution and the GEM approximation under the assumption that the first-order decay rate k is zero. At the location of the loading and everywhere downstream, the steady-state concentration will simply be the loading divided by the flow or 14.83 g/m³. Thus, no concentration gradients will be downstream of the loading and the numerical model should involve no discretization error. Upstream of the loading will be a concentration gradient as a result of the combined (and conflicting) effects of upstream dispersion and downstream advection.

We ran the GEM in steady-state mode under this conservative pollutant assumption using 51 compartments each of uniform length (8,000 m) and

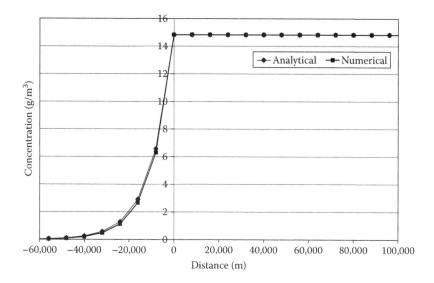

FIGURE 5.4 Discretization errors vis-á-vis concentration gradients.

volume (744,000 m³). The loading is in compartment 26, corresponding to $x = 0$. Compartments 1 and 51 are boundary compartments. We used the analytical solution corresponding to compartments 1 and 51 ($x = -200,000$ m and 200,000 m, respectively) to set the (fixed*) boundary conditions for the GEM simulation. A central difference ($\alpha = 0.5$) was used for the spatial differencing parameter. The portions of the analytical and numerical concentration profiles in proximity to the loading are shown in Figure 5.4, demonstrating the exactness of the numerical results downstream in the absence of a concentration gradient. Conversely, upstream of the loading where a concentration gradient is present, some (albeit minor) discretization error is apparent.

The reader can easily confirm the exact nature of the steady-state numerical solution in the absence of concentration gradients from Box 3.1 showing compartment i with four neighboring compartments. From that example, the mass balance equation at steady state is

$$0 = c_i\left[-E'_{i1} - Q_{i2} - E'_{i2} - Q_{i3} - E'_{i3} - Q_{i4} - E'_{i4}\right] + c_1\left(E'_{i1} + Q_{i1}\right)$$
$$+ c_2 E'_{i2} + c_3 E'_{i3} + c_4 E'_{i4} + W_i$$

If there are no concentration gradients, the dispersion has no driving force; thus, equivalently we can set the E terms to zero and the above equation can be solved for c_i as

$$c_i = \frac{c_1 Q_{i1} + W_i}{Q_{i2} + Q_{i3} + Q_{i4}}$$

---

* We also ran the simulation assuming a zero concentration gradient at the downstream boundary, to ensure that the fixed downstream boundary condition of 14.83 g/m³ (analytical solution) was not unfairly predetermining the concentration profile at $x > 0$. The two simulations gave identical results.

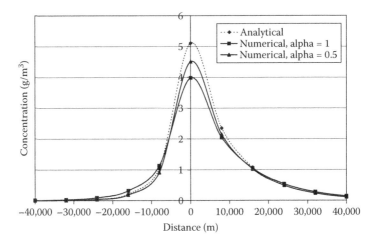

FIGURE 5.5   Numerical dispersion.

From a flow balance around compartment i, $Q_{i1} = Q_{i2} + Q_{i3} + Q_{i4}$ and

$$c_i = c_1 + \frac{W_i}{Q_{i2} + Q_{i3} + Q_{i4}}$$

The concentration in compartment i is simply the concentration advected from upstream plus the loading diluted by the flow, which is what we know the true solution to be in this simple case.

We next made several runs using $k = -0.25$/day to induce a concentration gradient downstream of the loading, and illustrate numerical dispersion. In Figure 5.5 the "Analytical" profile represents the analytical solution to this example with a dispersion coefficient of 1.3×10^7 m²/day. We next ran the GEM as described above (using $E = 1.3 \times 10^7$ m²/day and adding $k = -0.25$/day) and $\alpha = 1.0$. Finally, we made an otherwise identical run but using $\alpha = 0.5$. Those GEM results are shown in Figure 5.5 as "Alpha = 1" and "Alpha = 0.5." Neither result exactly matches the analytical, but the alpha = 0.5 solution is clearly better.

The difference between the two numerical solutions is the additional numerical dispersion induced by the backward difference. This dispersion tends to spread out the concentration profile by reducing the peak concentrations and increasing the concentrations elsewhere in the same manner as a "real" (physical) dispersion coefficient. The central difference does not involve numerical dispersion, and its errors relative to the analytical solution are the additional discretization types from ignoring Taylor series terms at and above third order. (Reducing the compartment lengths from 8,000 m would reduce these errors.) Admittedly, errors are errors and whether they are called numerical dispersions or something else seems semantic. Nonetheless, we are simply trying to distinguish the two and show the additional numerical dispersion error from the backward difference.

5.3.2 Positivity and Wiggle

While the use of a central difference instead of a backward difference may appear to be a panacea (no numerical dispersion and generally less total discretization error), one issue with a central difference does not occur with a backward difference. As described by Hall and Porsching (1990), when central differences are used for advective systems (where mass transport is predominantly due to advection rather than dispersion), solutions can become negative if the completely mixed compartments are too large, e.g., if Δx is too large in the one-dimensional examples used here.

Numerically, this problem can be traced to cases where the elements in the off-diagonal elements of the **A** matrix have the same sign as the main diagonal (i,i) elements (Hall and Porsching, 1990). Considering only transport terms for now (ignoring sources and sinks), from Equation (4.19) for concentration in compartment i, we can write the **A** matrix element of a neighboring compartment j as

$$\alpha_{ij}\overline{Q}_{ij} + E'_{ij}$$

if $\overline{Q}_{ij}$ is positive (flow from j into i) or

$$(1 - \alpha_{ij})\overline{Q}_{ij} + E'_{ij}$$

if $\overline{Q}_{ij}$ is negative (from i to j). ($\overline{Q}_{ij}$ is the water content-normalized flow, i.e., $\overline{Q}_{ij} = Q_{ij}/\theta_i$). The main diagonal elements of **A** will always be negative. Thus, to ensure positivity, we require for the off-diagonal elements that

$$\alpha_{ij}\overline{Q}_{ij} + E'_{ij} > 0 \text{ for } \overline{Q}_{ij} > 0 \qquad (5.38)$$

and

$$(1 - \alpha_{ij})\overline{Q}_{ij} + E'_{ij} > 0 \text{ for } \overline{Q}_{ij} < 0 \qquad (5.39)$$

We know that (5.38) will always be >0 because it applies to positive flows and the E' term is also positive. Thus, we can ignore those elements and (5.39) becomes our positivity criterion. Since the flows have a negative sign, we can write (5.39) in terms of the absolute value of flow as

$$-(1 - \alpha_{ij})\left|\overline{Q}_{ij}\right| + E'_{ij} > 0$$

or, rearranging, as

$$(1 - \alpha_{ij})\left|\overline{Q}_{ij}\right| < E'_{ij} \qquad (5.40)$$

How does compartment size affect this criterion? Let us substitute into (5.40)

$$\bar{U}_{ij} = \frac{|Q_{ij}|}{A_{ij}}$$

where $\bar{U}_{ij}$ is (positive-valued) water content-normalized velocity and A_{ij} is interfacial area and

$$E'_{ij} = \frac{E_{ij}A_{ij}}{L_{ij}}$$

where, as previously discussed, L_{ij} is the distance between the centroids of compartments i and j. Equation (5.40) can then be written as

$$(1-\alpha_{ij})\bar{U}_{ij}A_{ij} + \frac{E_{ij}A_{ij}}{L_{ij}} > 0$$

Solving for L_{ij} and canceling A_{ij} terms, we have a criterion for the maximum distance between compartments i and j (where flow is from i to j) as

$$L_{ij} < \frac{E_{ij}}{(1-\alpha_{ij})\bar{U}_{ij}} \qquad (5.41)$$

Note that the above criterion would require a L_{ij} for a forward difference $\alpha = 0$ that is half the length required for a central difference $\alpha = 0.5$, thus requiring a finer resolution. This disadvantage, combined with the fact that the forward difference also involves numerical dispersion, was the reason given earlier for dropping further consideration of the forward difference.

Often, one has already selected the compartment volumes and is then interested in the advective or dispersive conditions that may lead to negative concentrations. For this purpose, the above criterion is rearranged as

$$Pe = \frac{\bar{U}_{ij}L_{ij}}{E_{ij}} < \frac{1}{(1-\alpha_{ij})} \qquad (5.42)$$

where the quantity

$$\frac{\bar{U}_{ij}L_{ij}}{E_{ij}}$$

is termed the Peclet number Pe.

EXAMPLE 5.4

Illustration of Negative Concentrations

To illustrate this criterion using the above example with a Δx (i.e., L_{ij}) of 8,000 m, we ran the GEM using central differences for three different flow rates. The first flow corresponds to the flow rate at which the 8,000 m Δx just satisfies the above criterion ($Q = 301,000$ m³/day). We then ran the example at a flow 10% greater than that, and another run with the flow 100% greater. The results are shown in Figure 5.6. As can be seen, at flows greater than the critical flow, the concentration becomes negative just upstream of the loading.

The GEM automatically checks for positivity using Equation (5.41) for each compartment i (applied to all compartments j receiving flows *from i*) when central differences are used. If the criterion fails, the user is warned (in the diagnostic file wiggle.dng) but the run continues for reasons discussed below.

The positivity criterion (5.41) is *conservative*, i.e., it is a necessary but not sufficient condition for negativity. Violating it will not always result in negative concentrations. In regions where concentrations should be relatively high, a violation of this criterion is unlikely to numerically drive them to negative values. (In the preceding example, the criterion is violated along the entire length of the estuary simulated at flows above the critical flow. Nonetheless, only upstream of the loading, where concentrations will be near-zero anyway do negative concentrations occur.)

In addition, the criterion is expressed in terms of transport only and does not include the possibility of including source terms in the off-diagonal elements. These

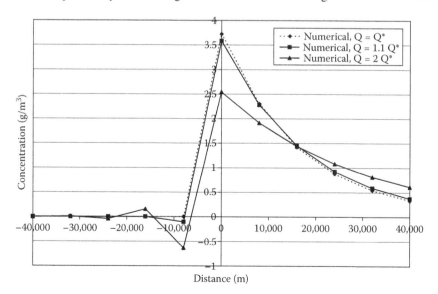

FIGURE 5.6 Negativity for central difference.

positive source terms* would add to the positivity of the off-diagonal elements. Furthermore, as discussed in Section 5.2, if a user specifies an effective mixing length between compartments i and j that is *less* than the distance between their centroids, the effect would be an increase of the E'_{ij} dispersion parameter by making the system less advection-dominated and therefore less susceptible to negative concentrations.

One might be justifiably curious whether a violation of the positivity or Peclet criterion is only a problem for possibly negative results. The answer is no, although negative concentrations are of obvious concern to modelers because they are physically impossible and that is why the criterion is often known as positivity. Indeed, more broadly, the criterion is sometimes called the wiggle criterion. When the criterion is violated, the solution will exhibit "wiggles," i.e., spatially decaying or growing oscillations of wavelength

$$\frac{L_{ij}}{1-\alpha_{ij}}$$

(Leonard, 1979) that may or may not involve negative values.

EXAMPLE 5.5

Illustration of Wiggles

To illustrate wiggles, we ran the previous example using a central difference, $Q = 3{,}000{,}000$ m³/day (to increase the Peclet number sufficiently to demonstrate significant wiggles) and an upstream boundary condition of 3.0 (to show wiggle without negativity). The resulting wiggles are shown in Figure 5.7.

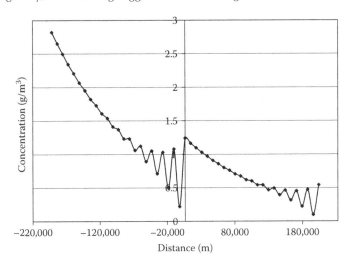

FIGURE 5.7   Wiggle for central difference.

---

* Note that negative sink terms are never included in off-diagonals; therefore they cannot non-conservatively affect the criterion.

In summary, for central differences ($\alpha_{ij} < 1.0$) the Peclet criterion provides an upper bound on compartment sizes that prevents both the possibility of negative concentrations and/or (numerically-induced) oscillations. Although most environmental modeling references use the term "positivity" to describe the criterion, we are going to use a broader definition and hereafter refer to the wiggle criterion.

5.3.3 CORRECTING FOR NUMERICAL DISPERSION AND RESULTING EQUALITY OF THE TWO DIFFERENCING METHODS

The disadvantage of the backward difference method cited above was that it introduces a numerical dispersion whereas the central difference does not. Conversely, the disadvantage of the central difference is that it is subject to wiggle and the backward difference is not. We are going to demonstrate that, *if a dispersion coefficient that corrects for the numerical dispersion* is used in the backward difference, then both differencing methods are mathematically equivalent, and both are subject to the wiggle criterion (and neither induces numerical dispersion).

To show this equivalency, first consider the mass balance equation for a compartment i subject only to transport processes and no external loadings. For simplicity, we assume a single spatial dimension and unity water content and no retardation. Thus, compartment i has an adjoining upstream compartment $i - 1$ and an adjoining downstream compartment $i + 1$. From Equation (4.19) we can write for a central difference,

$$\frac{d(V_i c_i)}{dt} = c_{i-1}(0.5 Q_{i,i-1} + E'_{i,i-1}) + c_i(0.5 Q_{i,i-1} - 0.5 Q_{i,i+1} - E'_{i,i-1} - E'_{i,i+1})$$
$$+ c_{i+1}(-0.5 Q_{i,i+1} + E'_{i,i+1}) \tag{5.43}$$

and for a backward difference,

$$\frac{d(V_i c_i)}{dt} = c_{i-1}(Q_{i,i-1} + E'_{i,i-1}) + c_i(-Q_{i,i+1} - E'_{i,i-1} - E'_{i,i+1}) + c_{i+1}(E'_{i,i+1}) \tag{5.44}$$

You will recall that

$$E'_{i,j} = \frac{E_{i,j} A_{i,j}}{L_{i,j}} \tag{3.7 repeated}$$

We identified three different dispersion coefficients E, namely the true physical dispersion coefficient, the numerical dispersion coefficient, and a dispersion coefficient actually used in the model. Let's keep these straight by calling them, respectively, E_p, E_n, and E_m. For the backward difference, because E_n is predictable and adds to E_p, we want to use in the model[†] a dispersion coefficient that negates E_n, i.e.,

---

[*] Recall that the numerical dispersion is *predictable*, i.e., $\bar{U} \Delta x / 2$ for a one-dimensional problem.

[†] Numerical issues of numerical dispersion, wiggle, and instability (discussed in Chapter 5) in the model are related to the dispersion coefficient *used in the model*, i.e., E_m. The model knows nothing about true physical dispersion; it knows only what the modeler specifies.

$$E_m = E_p - E_n \qquad (5.45)$$

Substituting (5.45) into (3.7) we have

$$E'_{m_{i,j}} = \frac{E_{m_{i,j}} A_{i,j}}{L_{i,j}} = \frac{(E_{p_{i,j}} - E_{n_{i,j}}) A_{i,j}}{L_{i,j}} \qquad (5.46)$$

Knowing that

$$E_n = \frac{U \Delta x}{2},$$

we can write (5.46) as (equating $L_{i,j}$ and Δx)

$$E'_{m_{i,j}} = \frac{\left(E_{p_{i,j}} - \dfrac{(U_{i,j} L_{i,j})}{2} \right) A_{i,j}}{L_{i,j}} = E'_{p_{i,j}} - 0.5 Q_{i,j} \qquad (5.47)$$

because $Q = UA$.

If we now substitute the dispersion coefficient used in the model (5.47) into the mass balance equation for the backward difference (5.44), all the E_n terms will cancel, the newly-introduced Q terms (Equation 5.47) modify the existing Q terms, and the resulting equation is *identical*[*] to that given above for the central difference (5.43). In summary, only one method will eliminate the numerical dispersion: a central difference. If you modify the backward difference dispersion coefficient to control for numerical dispersion, you are in fact using a central difference method, with its attendant advantage (no numerical dispersion) and disadvantage (subject to wiggle). To verify this equality, we reran the GEM on Example 5.3 using the backward difference but now specifying

$$E_m = E_p - \frac{U \Delta x}{2} = 1.3 \times 10^7 - \frac{(1,320)(8000)}{2} = 0.77 \times 10^7.$$

That run resulted in the exact profile shown in Figure 5.5 for $\alpha = 0.5$.

What does this rather startling result mean? Are we to believe that all the numerical modeling references that elaborate on the relative merits of backward versus central differencing schemes are simply missing the point and a choice among the methods is completely moot? No, because we do not *have* to correct the backward difference for E_n. If we do not, the backward difference remains a true backward difference and retains its wiggle-free advantage regardless of compartment sizes. In some situations, this feature may be more desirable than minimizing numerical dispersion.

---

[*] With the understanding that the E in the central difference equation is now our true physical E_p, that is, we would use $E_m = E_p$ in the central difference equation because there is no E_n.

Note: From this point forward, for simplicity we will often refer to the $\alpha = 1$, backward difference as a "back space" (BS) method and the $\alpha = 0.5$, central difference as a "centered space" (CS) method.

One situation where the BS method *not* corrected for numerical dispersion has a major advantage is pure advection ($E_p = 0$). Using the E-unmodified BS method is appropriate in this situation *not* because we are willing to live with the numerical dispersion (apparently, none is involved) but because: (1) using a central difference (CS or E-modified BS) is unconditionally wiggly, and (2) a central difference method involves a downstream boundary condition that may be problematic.

From the wiggle criterion (5.41) you can see that if a zero dispersion coefficient is used, the maximum length for a central difference method (or backward difference with $E_m = 0 - E_n$) to avoid wiggle is 0, which is not very helpful. In addition, a central difference method would require specification of a downstream boundary condition.[*] Recall that adjoining compartment concentrations come into play in the advection terms for $\alpha = 0.5$ [e.g., c_{i+1} in Equation (5.43)]. This means that a downstream boundary condition will be required in addition to an upstream boundary condition.

This is not only problematic in practice, but is an affront to our intuition in that we know that purely advective systems do not look downstream to determine what concentrations should be. (By definition, pure advection means all mass transport is from upstream so nothing downstream of compartment i can influence concentrations in i.) Indeed, knowing a downstream boundary condition in a purely advective system is no less problematic than knowing the concentrations in the regions for which the model is being developed, because such a boundary condition would have to be predicted from upstream concentrations.

EXAMPLE 5.6

Central Differences and Pure Advective Systems

Figure 5.8 illustrates the problem inherent with using $\alpha = 0.5$ for a purely advective system. The example is the previous Thomann and Mueller advective/dispersive system but with $E_m = E_p = 0$. Where we previously used $\Delta x = 8{,}000$ m, we decreased that for purposes of this example by almost an order of magnitude or $\Delta x = 1{,}000$ m to further illustrate that the central difference wiggle problem persists regardless of compartment size for purely advective systems.

The compartment volumes became 93,000 m³. The flow rate used was the previous critical flow for no wiggle (for $\Delta x = 8{,}000$ m) of 301,000 m³/day. Three solutions are presented, including the analytical solution to this problem.[†] The two

---

[*] You could finesse this requirement by using a linear or zero gradient downstream boundary condition, but the problem of unconditional wiggle remains.

[†] The analytical solution for the pure advection case is

$$c(x) = \frac{W}{Q} \exp(-kx/U)$$

for $x \geq 0$ and 0 everywhere upstream ($x < 0$).

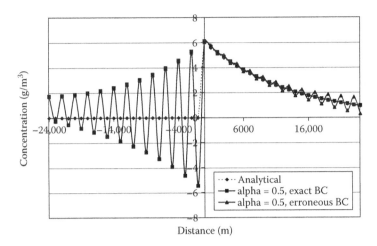

FIGURE 5.8   Wiggle inherent with CS methods on pure advection example.

GEM solutions are: (1) a central difference $\alpha = 0.5$ with $E_m = 0$ and an *exact* downstream boundary condition given by the analytical solution (0.87), and (2) another central difference $\alpha = 0.5$ with $E_m = 0$ and a slightly *erroneous* downstream boundary condition of 1.6 g/m³.

As can be seen in Figure 5.8, although the central difference solution[*] with the known exact downstream boundary condition is good at the loading and downstream, it suffers from wild oscillations and even negative concentrations upstream. In fact, however, even if one were willing to live with the upstream wiggle, it is highly unlikely that this solution would be realizable in an application lacking an analytical solution (which is why you would build a numerical model in the first place) because the exact downstream boundary condition would in general be unknown—as we argued previously for purely advective systems. You would have to guess at the downstream boundary condition. Suppose the guess were only off by a factor of 2, e.g., suppose a value of 1.6 was assumed for a downstream boundary condition. That run with the 1.6 downstream boundary condition is also shown in Figure 5.8 and suffers from wiggle throughout the spatial domain, i.e., not only the wild upstream oscillations, but downstream as well.

To reiterate, this wiggle problem persists in the pure advection case for a central difference solution (or an equivalently E_n-modified backward difference) regardless of how small we make Δx. We also ran this example using a much smaller $\Delta x = 80$ m and the exact downstream boundary condition. Interestingly, the central

---

[*] To further illustrate the equivalence between a central difference and a backward difference with $E_m = -E_n$, we also ran this as a backward difference with $E_m = -E_n = -(3237)(1000)/2 = -1.618 \times 10^6$. The same upstream and exact downstream boundary conditions were as used for the central difference. Again, the two results are identical. Note that we used a negative dispersion coefficient E_m to offset E_n. We are NOT simulating negative dispersion. Negative (real) dispersion is physically impossible; it would involve molecules moving from low concentrations to high concentrations on average. We are merely offsetting positive numerical dispersion. Any E_m more negative than this value, however, would be an attempt to simulate negative dispersion and not appropriate.

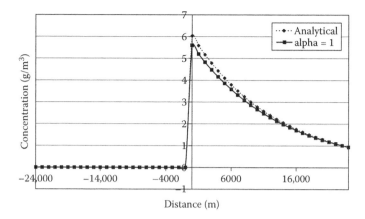

FIGURE 5.9   Pure advection example using BS methods.

difference solution was even worse than the result for $\Delta x = 1,000$ m. Wild oscillations occurred throughout the domain, even downstream of the loading. Finally, Figure 5.9 shows a backward difference ($\alpha = 1.0$) simulation with $E_m = 0$ and $\Delta x = 1,000$ m. Despite some loss of resolution at the peak concentration, *wiggle is not an issue*. Therefore, in the pure advection case, our best choice would be true backward difference ($\alpha = 1.0$ and $E_m = 0$) with appropriately small compartment volumes (at least in the vicinity of sharp gradients; see following discussion) to mitigate discretization error.

A brief aside is in order regarding numerical dispersion when using the E-unmodified BS method for a pure advection problem. Interestingly, and despite the numerical dispersion indicated by the Taylor Series analysis (Section 5.3.1), it appears that no numerical dispersion was induced by the E-unmodified BS method *for the pure advection problem*. First, comparing the slopes of the BS results and the analytical results in Figure 5.9 downstream of the loading, they are very similar. Numerical dispersion would produce different slopes. Perhaps the example is not advective enough to generate noticeable numerical dispersion. (Recall that numerical dispersion increases with advective flow rate.) To address that question, we quadrupled the flow rate from 301,000 to 1,200,000 m³/day and reran the problem. Again, the BS results were in excellent agreement with the analytical results with essentially no differences in their slopes.

As a less empirical argument for the absence of numerical dispersion, Chapra (1997), in the context of discussing implicit methods for solving dynamic problems (our Chapter 6), states that implicit methods (e.g., BTBS in Chapter 6) may generate *intermediate* numerical dispersion but, as one runs those solutions out to a steady-state solution, the numerical dispersion drops out. Chapra cites that as a reason that implicit methods that are unconditionally stable, thus allowing large time steps, are often used to solve steady-state problems.

To check this claim, we ran this pure advection example out to steady state using the GEM's BTBS dynamic method (with a 1-day time step and 500 time steps) and the results were identical to the steady-state BS results. Finally, we make the argument in Chapter 6 that the explicit FTBS method is well suited to the pure advection problem (no intermediate or other numerical dispersion) if one uses the maximally stable time step. Running the GEM's FTBS option at the maximum time step again out to steady state also exactly replicates the steady-state BS results.[*]

Based on these considerations, we conclude that the E-unmodified BS method for the pure advection problem not only is wiggle-free but does not result in numerical dispersion despite the Taylor series analysis. (We don't see where the Taylor series analysis is irrelevant when $E_p = 0$, but perhaps others will.)

Relaxing now our assumption of pure advection, as systems become increasingly dispersive and no wiggle *can* be assured with $\alpha = 0.5$ (by suitably selecting compartment sizes) and downstream boundary conditions are not problematic, it becomes more advantageous to balance the trade-off between numerical dispersion and compartment sizes by moving toward central differences, i.e., $0.5 \leq \alpha_{i,j} < 1.0$. There is probably an optimal strategy for determining the $0.5 \leq \alpha_{i,j} < 1.0$ values that will maximize L_{ij} while ensuring no wiggle and satisfying constraints on acceptable numerical dispersion (that exercise is left to others).

Interestingly, at the opposite extreme of pure dispersion ($Q = 0$), the choice of α becomes irrelevant because α affects only the advection terms in the model. Furthermore, there will be no numerical dispersion; i.e., from (5.26) if $\bar{U}$ is 0, then E_n is 0. Concentrations will also be unconditionally wiggle-free; the off-diagonal elements of matrix $\mathbf{A}$ are guaranteed to be positive if flows are 0. Because we are still dealing with a numerical approximation to the dispersion (second derivative) terms, discretization error will occur. Therefore, smaller compartments in the vicinity of concentration gradients are still needed to minimize discretization errors.

To summarize, if $\alpha = 0.5$ and $E_m = E_p$, we have no numerical dispersion, but must be concerned about selecting compartment sizes that ensure no wiggle. Conversely, if $\alpha = 1$ and $E_m = E_p$, no wiggle is assured, but numerical dispersion is an issue (except when $E_p = 0$). If $\alpha = 1$ and $E_m = E_p - E_n$, then, as we have seen, we are effectively using a central difference. (If $\alpha = 0$ is used, as discussed previously, both numerical dispersion and wiggle are issues and we will not further consider it.)

5.3.4   QUANTIFICATION OF TOTAL DISCRETIZATION ERROR

Before we discussed the implications of choice of a spatial differencing parameter on different types of numerical errors, we introduced total discretization error and demonstrated that it occurs where concentration gradients are present. What is desirable for a modeler trying to determine appropriate compartment sizes is a method of selecting sizes so that some acceptably small error is attained, given the trade-off between decreasing error (small compartments) on one hand and decreasing system complexity and run times (large compartments) on the other.

In response to this need, Mueller (1976, as described in Thomann and Mueller, 1987, and Chapra, 1997) developed relationships between discretization errors and

---

[*] Running the FTBS option at time steps *less* than the maximum stable (that involves at least intermediate numerical dispersion as discussed in Chapter 5) also gives identical results. Chapra's statement that numerical dispersion in dynamic simulations drops out as steady state is approached for implicit methods apparently also extends to explicit methods.

compartment sizes (Δx) for one-dimensional, advective and dispersive systems with first-order decay and constant coefficients, as described by Equation (5.20). No retardation and unity water content were implicitly assumed. These relationships apply to regions of maximum concentration—i.e., where loadings occur—and where discretization errors would be expected to be of most concern because gradients will be sharpest there. These relationships were also developed assuming a backward difference scheme ($\alpha = 1$); therefore, numerical dispersion is included in the error.

For a desired fractional relative error ε between the numerical and analytical models, the distance between compartment centroids at the location of a load should be

$$\Delta x = \frac{U}{k}\sqrt{\frac{1+\eta}{(1+\varepsilon)^2}} - \eta - 1 \quad \text{for } \eta < 1 \quad \text{(advective system)} \qquad (5.48)$$

and

$$\Delta x = \sqrt{\frac{4E\left(\sqrt{\frac{1+\frac{1}{\eta}}{(1+\varepsilon)^2}} - 1 - \frac{1}{\sqrt{\eta}}\right)^2}{k}} \qquad \text{for } \eta > 1 \quad \text{(dispersive system)} \qquad (5.49)$$

where η is the so-called "estuary number"

$$\eta = \frac{4kE}{U^2} \qquad (5.50)$$

EXAMPLE 5.7

Calculating Desired Relative Error

To illustrate, our previous Example 5.3 ($E_p = 1.3 \times 10^7$ m^2/day, $U = 1,320$ m/day, $k = 0.25$/day) has an estuary number of

$$\eta = \frac{(4)(0.25)(1.3 \times 10^7)}{1320^2} = 7.48$$

so that the criterion of (5.49) applies. At $x = 0$ (loading point), the analytical solution (5.21) gives a peak concentration of 5.09 g/m^3. Our numerical solution (backward difference, $E = 1.3 \times 10^7$ m^2/day, $\Delta x = 8000$ m) gave a peak concentration of 3.99 g/m^3 for a relative error $\varepsilon = (3.99 - 5.09)/5.09 = -0.22$. Substituting that ε into Equation (5.49) along with the other parameters, the calculated Δx is 8,047 m, in close agreement with the Δx used in the numerical solution.

REFERENCES

Chapra, S.C. 1997. *Surface Water Quality Modeling*, preliminary ed. New York: McGraw-Hill.

Dennis, J.E., Jr. and Schnabel, R.B. 1983. *Numerical Methods for Unconstrained Optimization and Nonlinear Equations*. Philadelphia: Society for Industrial and Applied Mathematics.

Hall, C.A. and Porshling, T.A. 1990. *Numerical Analysis of Partial Differential Equations*. Cambridge: Cambridge University Press.

Hoffman, J.D. 1992. *Numerical Methods for Engineers and Scientists*. New York: McGraw-Hill.

Hornbeck, R.W. 1975. *Numerical Methods*. New York: Quantum Publishers.

Matthews, J.H. 1992. *Numerical Methods for Mathematics, Science, and Engineering*, 2nd ed. New York: Prentice Hall.

Mueller, J.A. 1976. *Accuracy of Steady State Finite Difference Solutions*. Technical Memorandum. Hydroscience, Inc.

Press, W.H., Flannery, B.P., Teukolsky, W.T. et al. 1986. *Numerical Recipes: The Art of Scientific Computing*. Cambridge: Cambridge University Press.

Rosenbrock, H.H. 1960. An automatic method for finding the greatest or least value of a function. *Computer J.*, 3: 174.

Strang, G. 1986. *Introduction to Applied Mathematics*. Wellesley, MA: Wellesley-Cambridge Press.

Thomann, R.V. 1972. *Systems Analysis and Water Quality Management*. New York: McGraw-Hill.

Thomann, R.V. and Mueller, J.A. 1987. *Principles of Surface Water Quality Modeling and Control*. New York: Harper & Row.

6 Solution Techniques for Dynamic Problems

6.1 INTRODUCTION

Recall that our generalized system of equations can be written in matrix notation for problems with linear source and sink terms as

$$\frac{d(\mathbf{RV}\underline{c})}{dt} = \mathbf{A}\underline{c} + \underline{b}$$

(4.20 repeated)

and for nonlinear systems as

$$\frac{d(\mathbf{R}(\underline{c})\mathbf{V}\underline{c})}{dt} = \mathbf{A}\underline{c} + \underline{f}(\underline{c}) + \underline{b}$$

(4.22 repeated)

In Chapter 5, we discussed the steady-state solutions to these equations. Here we discuss how to solve these dynamic equations in time.

This chapter presents four GEM options for numerically advancing a solution in time from some initial condition. Although many dynamic numerical solution techniques are available, these four were selected because each has some significant advantage, and/or is illustrative of a different and interesting approach. Steve Chapra's excellent overview of these four methods (1997) strongly motivated their inclusion in the GEM. The reader is encouraged to augment the presentation herein with Chapra's overview.

The first two methods (forward time Euler and centered time MacCormack) are explicit in that they march the solution from a current time step t to the next time step $t + 1$ by using only information already available at time step t. Thus, we can readily advance the solution to time step $t + 1$ because all the needed information is available and we simply solve for the new concentrations explicitly. Explicit methods do not require solving systems of simultaneous equations, and are therefore mathematically indifferent to whether the mass balance equations are linear or have the $\underline{f}(\underline{c})$ type of nonlinearity. However, for $\mathbf{R}(\underline{c})$-type nonlinearities, the explicit methods cannot be used because the unknown concentrations to be solved at a time step cannot be isolated linearly on the left side of the equation and an iterative method is needed. The quasi-Newton's method (see Chapter 5) is used by the GEM for iterating on nonlinear systems under steady-state or dynamic modes.

In contrast, the last two methods (back time and centered time or Crank–Nicolson) are implicit. They march the solution forward in time using information both from the prior time step t *and* the new time step $t + 1$. Because time step $t + 1$ information is used, it is necessary to solve a system of simultaneous algebraic equations at each time step, i.e., the solution is only known implicitly by solving these equations. For linear systems, linear algebra methods are used. For nonlinear systems, the quasi-Newton method is used.

The GEM solution options use the forward time (FT), backward time (BT), and centered time (CT) descriptions that need clarification before we go further. Recall that in the steady-state case, our spatial discretization scheme for the advective term arose from the need to express the concentration at the interface between compartments i and j ($c_{i,j}$) as some function of concentrations c_i and c_j within the adjoining compartments.

Whether the discretization was backward ($\alpha = 1$), forward ($\alpha = 0$), or central ($\alpha = 0.5$) had meaning only for the advective term. We were able to make a discretization for the advective term, without reference to the other terms in the equation, because those other terms were already implicitly discretized in space by the compartment approach. In contrast, in the dynamic case our temporal discretization approach must consider the *entire* equation, because we are advancing all terms in time and none of those terms was previously temporally discretized. For example, consider a simple time-dependent differential equation

$$\frac{dy}{dt} = f(y(t)) \tag{6.1}$$

where the change in the dependent variable y over time is given by some function f of y. Because y is a function of time, f is also a function of time. If we make, say, a forward temporal difference to the derivative only (as we did in the steady-state case for the spatial derivative),

$$\frac{y^{t+1} - y^t}{\Delta t} = f(y(t)) \tag{6.2}$$

so that we can try to advance the solution from time t to time $t + 1$ as

$$y^{t+1} = y^t + \Delta t f(y(t)) \tag{6.3}$$

That doesn't get us anywhere because we have not specified at what discrete time $f(y(t))$ is to be evaluated. We could complete our temporal discretization scheme by evaluating f at time t to result in

$$y^{t+1} = y^t + \Delta t f(y(t))^t \tag{6.4}$$

which is called a forward time (FT) approach. As previously mentioned, this is an explicit method because, as can be seen above, we can solve explicitly for y^{t+1} as all terms on the right side of the equation are known (were previously evaluated) at time t. Alternatively, we could choose to complete our temporal discretization scheme by evaluating f at time $t + 1$ to result in

$$y^{t+1} = y^t + \Delta t f(y(t))^{t+1} \tag{6.5}$$

which is called a backward time (BT) approach. This an implicit method, because now we *cannot* solve explicitly for y^{t+1} and must do so implicitly. In principle, we could also select a central difference temporal discretization approach or

$$y^{t+1} = y^{t-1} + 2\Delta t f(y(t))^t \tag{6.6}$$

Similar to the use of a central difference for the spatial derivative, the advantage of this approach is a reduction in discretization error (and, notably, zero temporal numerical dispersion). Interestingly, however, this approach has been shown to be unconditionally unstable (stability is discussed later) according to Remson et al. (1971) and is to be avoided. The MacCormack and Crank–Nicolson methods discussed later avoid this limitation. The MacCormack method approximates a central temporal difference while the Crank–Nicolson method is a fully central difference that does not suffer from unconditional instability.

Numerical methods for solving partial differential equations are also characterized by how the space domain is discretized. As discussed previously in the context of the GEM, this is parameter α in the advective transport term; $\alpha = 1$ is a backward spatial difference; $\alpha = 0$ is a forward spatial difference; $\alpha = 0.5$ is a central difference. Thus, a method based on, say, a central temporal difference and a central spatial difference would be described as a centered time, centered space (CTCS) method. With one exception (the MacCormack method), we do not force the value of α into the following options, but rather let the user choose it as desired. The rationale for choosing α was discussed in Section 5.3 and that rationale is still relevant for dynamic problems.

The four GEM options discussed in this chapter all use two time levels in their discretization schemes. Unsurprisingly, these are known as two-level schemes (Vestegaard, 1989). Although not commonly used in environmental modeling, the reader should be aware that other schemes that use three or more levels are possible.

Chapter 5 included an extensive discussion of numerical errors associated with steady-state problems, i.e., numerical dispersion, positivity, wiggle, and total discretization errors that remain issues for dynamic problems. In addition, the extension of the numerical solution into the time domain introduces an additional concern— stability. Similar to the wiggle oscillations in space discussed previously but now wiggling in time, an unstable solution causes numerical errors to accumulate and grow at each time step. The result is a rapidly oscillating solution in which concentrations increase and decrease without bound until the solution numerically "blows up" and a computer run-time overflow error occurs.

The development of all four GEM dynamic options includes a discussion of their stability characteristics. In addition, because time discretization may introduce a temporal component to numerical dispersion (in addition to the spatial component), numerical dispersion characteristics of each method are also discussed. The four options are illustrated and compared by examples.

6.2 EXPLICIT FORWARD TIME (EULER) METHOD

6.2.1 METHODOLOGY

This is the simplest method possible based on ease of understanding, mathematical simplicity, minimal coding, and robustness to linear and some nonlinear problems. If you had been the first numerical analyst to try to solve a differential equation following the discovery of the concept of a finite difference, this is the method you would have developed. (That pioneer was Leonhard Euler, the pre-eminent mathematician of the eighteenth century.) The method is so simple that, for small problems (few compartments and chemicals), one could implement it for a few time steps by hand calculations.

Thus, it is an excellent method to begin implementation of the GEM to a new problem because one could relatively easily check GEM results with these hand calculations. After successfully running the GEM with this method, confidence would be gained that the problem is set up correctly and the user could then move on directly to more sophisticated GEM solution options.

Because of its simplicity, the method has a major computational advantage. For linear and/or nonlinear problems of the $\underline{\mathbf{f}}(\underline{\mathbf{c}})$-type nonlinearity, it does not involve solving systems of equations. The chief disadvantage is that the method suffers from relatively restrictive conditions on the size of the time step that can be used while still maintaining stability. In addition, numerical dispersion can be significant unless properly corrected for.

The forward time (FT) method can be implemented as a FTBS, FTFS, or FTCS method, depending on the user's choice of α. We previously mentioned in the steady-state discussion that there is little to recommend the FS (α = 0) option (with one exception; see the MacCormack method description). That applies even more to dynamic problems because the FTFS method is unconditionally unstable (even for very small time steps) as reported by Vestegaard (1989). We confirmed that using the GEM.

Let the current time step be t (the solution has been advanced to time t). The FT Euler approximation to our general equation (4.22) is then

$$\frac{(\mathbf{R}(\underline{\mathbf{c}})\mathbf{V}\underline{\mathbf{c}})^{t+1} - (\mathbf{R}(\underline{\mathbf{c}})\mathbf{V}\underline{\mathbf{c}})^{t}}{\Delta t} = (\mathbf{A}\underline{\mathbf{c}} + \underline{\mathbf{f}}(\underline{\mathbf{c}}) + \underline{\mathbf{b}})^{t} \tag{6.7}$$

which can be expressed explicitly because all information at time t is available to calculate the NEQN × 1 $\mathbf{R}(\underline{\mathbf{c}})\mathbf{V}\underline{\mathbf{c}}$ vector at $t + 1$ as a function of known terms from the previous time step t as

$$(\mathbf{R}(\underline{\mathbf{c}})\mathbf{V}\underline{\mathbf{c}})^{t+1} = (\mathbf{R}(\underline{\mathbf{c}})\mathbf{V}\underline{\mathbf{c}})^{t} + \Delta t (\mathbf{A}\underline{\mathbf{c}} + \underline{\mathbf{f}}(\underline{\mathbf{c}}) + \underline{\mathbf{b}})^{t} \tag{6.8}$$

For linear systems, the desired solution for the constituent concentrations at $t + 1$ can be easily obtained by left-multiplying both sides of Equation (6.8) by the inverse of $(\mathbf{RV})^{t+1}$, i.e., $((\mathbf{RV})^{-1})^{t+1}$ to obtain

$$\underline{c}^{t+1} = ((\mathbf{RV})^{-1})^{t+1}[(\mathbf{RV}\underline{c})^t + \Delta t(\mathbf{A}\underline{c} + \underline{b})^t] \tag{6.9}$$

Because both $\mathbf{R}$ and $\mathbf{V}$ are both main diagonal matrices, the $(\mathbf{RV})^{-1}$ inverse of their product has a particularly simple structure consisting simply of the reciprocal of the product of the individual compartment volumes times R_i (Equation 4.18) on the main diagonal, i.e., main diagonal element i,i of $(\mathbf{RV})^{-1}$ is $\theta_i/[(V_i)(\Theta_i + BD_iK_{d_k})]$ with zeros elsewhere.

We are sensitive to the lack of extensive familiarity with matrix notation for many readers. At this juncture, some eyes may glaze over as they contemplate Equation (6.9) and wonder how to use a complex matrix equation such as that to find concentrations at time $t + 1$. Although we show (6.9) as the solution to a *system* of equations, one does not need a system approach to solve for concentrations. The matrix (system) representation is used merely for efficiency of presentation. Let's consider an arbitrary, ith row of the system of equations given by (6.9):

$$c_i^{t+1} = \frac{1}{R_{i,i}^{t+1}V_{i,i}^{t+1}}\left[R_{i,i}^t V_{i,i}^t c_i^t + \Delta t\left(\sum_j A_{i,j}^t c_i^t + b_i^t \right) \right] \tag{6.10}$$

All one has to do to find c_i^{t+1} is solve the single Equation (6.10). All terms on the right side are known because they are parameters or because the concentrations at time t were determined previously. One simply solves Equation (6.10) for all compartments i using boundary and initial conditions as appropriate.

It is often convenient to visualize a particular combination of a temporal and spatial discretization scheme by means of a stencil of grid points in space and time and determine how at any point in space and time for which an updated solution is needed the nearby points are used in the solution. For example, if we denote the discrete time points as $t - 1, t, t + 1$, etc. and the discrete space points as $i - 1, i, i + 1$, etc., the FTCS[*] method would be illustrated as shown in Figure 6.1

The spatial dimension in Figure 6.1 is shown as if it is one-dimensional. For multi-dimensional problems, it should be understood that grid point $i -1$ actually denotes *all* compartments j adjacent to compartment i that contribute advective flow to compartment i. Similarly, grid point $i + 1$ actually denotes *all* compartments j that receive advective flow from compartment i.

Assuming all concentrations are known at some time t (and all previous times), the unknown concentration at spatial point (compartment) i for time $t + 1$ is calculated given[†] the concentrations at time t in compartments $i + 1$ and $i - 1$. After $c(t + 1,i)$ is

---

[*] For the FTBS method, the figure would *not* include the directed arrow from node $i + 1$ at time t.
[†] The initial conditions get us started.

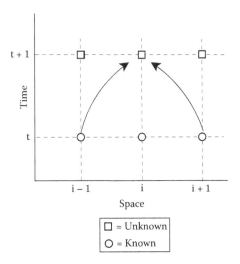

FIGURE 6.1 Time and space stencil for FTCS method.

calculated from $c(t, i-1)$ and $c(t, i+1)$, we can proceed across the other compartments at $t+1$ and fill in all the values in like manner. After we calculate all compartment concentrations at time $t+1$, we advance to the next time step ($t+2$, where all values at $t+1$ are now known) and repeat. When we get to the edges of the space rows, e.g., $i = 1$ or $i = $ NEQN, the $i-1$ or $i+1$ value is known from boundary conditions.

For the nonlinear case where $\mathbf{R}$ is not a function of $\underline{c}$, the Euler update is analogous to the linear case and significantly does not involve solving nonlinear systems of equations. That update is

$$\underline{c}^{t+1} = ((\mathbf{RV})^{-1})^{t+1}\{(\mathbf{RV}\underline{c})^t + \Delta t[\mathbf{A}\underline{c} + \underline{f}(\underline{c}) + \underline{b}]^t\} \qquad (6.11)$$

It is interesting to note that the Euler approach could be used in this situation, i.e., nonlinear $\underline{f}(\underline{c})$ and linear $\mathbf{R}$, to find a *steady-state* solution to a nonlinear problem without having to resort to iterative nonlinear methods by simply running the Euler update (Equation 6.11) out to steady state.

For the more general nonlinear problem where $\mathbf{R}(\underline{c})$ involves nonlinearities, Equation (6.8) would be expressed in implicit form (set equal to 0) suitable to an iterative approach such as Newton's method for finding roots of nonlinear equations as

$$\underline{g}(\underline{c}) = [\mathbf{R}(\underline{c})\mathbf{V}\underline{c}]^{t+1} - [\mathbf{R}(\underline{c})\mathbf{V}]^t - \Delta t[\mathbf{A}\underline{c} + \underline{f}(\underline{c}) + \underline{b}]^t = \underline{0} \qquad (6.12)$$

where $\underline{g}(\underline{c})$ is the implicit function (see Section 5.2.1) and $\underline{0}$ is the NEQN × 1 vector of $\bar{0}$s. Because Equation (6.12) has been set to $\underline{0}$, finding the unknowns of this nonlinear system (again, the elements of $\underline{c}^{t+1}$) is equivalent to finding the NEQN roots of the system of equations.[*] Note that the elements of the $\underline{c}$ vector at time $t+1$

---

[*] One could solve each nonlinear equation individually, but it is much more practical to use a systems approach and iterate on all equations simultaneously.

are the unknowns to be determined by the Newton algorithm for time step $t + 1$. Concentrations up to time t are known and are simply numerical data in the context of the Newton algorithm at time $t + 1$. Thus, the last two terms on the left side of Equation 6.12 (terms with superscript t) contain only data, not unknowns. The i,jth element of the Jacobian matrix is

$$\frac{\partial g_i}{\partial c_j} = \frac{\partial [\mathbf{R(c)Vc}]_i^{t+1}}{\partial c_j} = V_{ij} \frac{\partial [R_{ij}(\mathbf{c})c_i]^{t+1}}{\partial c_j} \tag{6.13}$$

where

$$\frac{\partial [R_{ij}(\mathbf{c})c_i]^{t+1}}{\partial c_j} = \left[R_{ij}(\mathbf{c}) + c_i \frac{\partial R_{ij}(\mathbf{c})}{\partial c_j} \right]^{t+1} \quad \text{if, } i = j \tag{6.14a}$$

$$\frac{\partial [R_{ij}(\mathbf{c})c_i]^{t+1}}{\partial c_j} = 0 \quad \text{if, } i \neq j \tag{6.14b}$$

$R_{ij}(\mathbf{c})$ and V_{ij} denote the i,jth elements of main diagonal matrices $\mathbf{R}$ and $\mathbf{V}$, respectively. Because $\mathbf{R}$ and $\mathbf{V}$ are main diagonal matrices, these elements are 0 if i is not equal to j. If analytical derivatives are used, the user supplies the partial derivatives of $R_{ij}(\mathbf{c})$ with respect to c_j, i.e., $\partial R_{ij}(\mathbf{c})/\partial c_j$ for all i and j.

6.2.2 STABILITY

As previously noted, the chief disadvantages of the Euler method are the relatively small time steps and concomitant increased computational burden required to prevent numerical instability. This section presents the criteria for detecting instability for both the linear case and the special nonlinear case where the $\mathbf{R}$ matrix is not a function of $\mathbf{c}$.

Recall that the generalized linear model using the Euler approach is

$$\underline{c}^{t+1} = ((\mathbf{RV})^{-1})^{t+1}\{(\mathbf{RVc})^t + \Delta t[\mathbf{Ac} + \underline{f}(\underline{c}) + \underline{b}]^t\} \tag{6.11 repeated}$$

For linear systems, the ith row of this system of equations can be written as

$$c_i^{t+1} = \frac{1}{R_{i,i}^{t+1}V_{i,i}^{t+1}} \left[R_{i,i}^t V_{i,i}^t c_i^t + \Delta t \left(\sum_j A_{i,j}^t c_i^t + b_i^t \right) \right] \tag{6.10 repeated}$$

and, grouping coefficients of c_i^t, can be written as

$$c_i^{t+1} = \frac{1}{R_{i,i}^{t+1}V_{i,i}^{t+1}} \left[\left(R_{i,i}^t V_{i,i}^t + \Delta t A_{i,i}^t \right) c_i^t + \Delta t \left(\sum_{j \neq i} A_{i,j}^t c_j^t + b_i^t \right) \right] \tag{6.15}$$

It can be shown (Carnahan et al., 1969) that the solution is stable if the main diagonal elements of the system of Equations (6.11), i.e., the coefficients of c_i^t in Equation (6.15), are positive. This stability criterion is

$$\frac{1}{R_{i,i}^{t+1}V_{i,i}^{t+1}}(R_{i,i}^t V_{i,i}^t + \Delta t A_{i,i}^t) > 0 \text{ for all } i \text{ and } t$$

and, because the denominator will always be positive and can be ignored, can be more simply expressed as

$$R_{i,i}^t V_{i,i}^t + \Delta t A_{i,i}^t > 0 \qquad (6.16)$$

This criterion can be shown to have a physical interpretation as discussed in Box 6.1.

BOX 6.1   COURANT CONDITION

For backward or central spatial discretizations, it can be shown that the value of the A_{ij}^t term in (6.16) is less than or equal to zero in the GEM structure. It is zero for a central difference and no dispersion; otherwise it is negative. Recall our sign convention for advective flows. Thus, we can equivalently write (6.16) using the absolute value of $A_{i,i}^t$ as

$$R_{i,i}^t V_{i,i}^t - \Delta t |A_{i,i}^t| > 0 \qquad (6.17)$$

which can be solved for Δt as

$$\Delta t < \frac{R_{i,i}^t V_{i,i}^t}{|A_{i,i}^t|} \qquad (6.18)$$

which has a physical meaning. Recall that the elements of the **A** matrix have units of flow (m^3/day). Elements of the retardation matrix **R** are unitless and volume divided by flow is a residence time. Thus the $R_{i,i}^t V_{i,i}^t / |A_{i,i}^t|$ term is effectively a chemical residence time (the time a chemical takes to be transported through compartment i, taking into account transport, source, sink, and retardation phenomena). Thus, the physical meaning of the stability criterion is that the numerical time step must be less than the chemical residence time.

Consider the case for a conservative chemical subject only to advective transport (no retardation), i.e., $k = E = 0$ and water content and $R = 1$ spatially and temporally constant parameters in a one-dimensional system and a backward difference. The diagonal elements of **A** are simply the (absolute value of) advective flow, Q, and (6.18) can be written as

$$\Delta t < \frac{V}{Q} \qquad (6.19)$$

Substituting for $Q = U*A$ where U is velocity and A is cross-sectional area and $V = A*\Delta x$ where Δx is compartment length (in the flow direction), we can express (6.19) as

$$\Delta t < \frac{\Delta x}{U} \tag{6.20}$$

and see that the time step must be less than the time taken for advective transport to traverse the compartment length. In other words, the numerical time step cannot go faster than the system, or "get ahead of itself." Equation (6.20) is commonly called the Courant condition.

For nonlinear systems with $\underline{f}(\underline{c})$, but $\mathbf{R} \neq \mathbf{R}(\underline{c})$, we extended the above stability criterion (6.16) for the GEM[*] to

$$R_{i,i}^t V_{i,i}^t + \Delta t (A_{i,i}^t + f_i(\underline{c})^t) > 0 \quad \text{for all } i \text{ and } t \tag{6.21}$$

Given a user-supplied time step, the GEM monitors the stability criterion each time step (the criterion can change temporally for nonlinear problems or linear problems with time-varying parameters) and outputs to a diagnostic file any time steps at which the criterion is violated. For linear problems, the most limiting time step over the duration of the simulation is determined and the critical time step that will satisfy stability is also reported to the diagnostic file. The user would then use this critical time step in a subsequent run.[†] For nonlinear problems, this critical time step cannot be determined because the criterion also involves $f_i(\underline{c})$, which varies with changes in $\underline{c}$ from run to run. Instead, the user would determine the critical time step by trial and error. (Suggestion: halve the time step until stability is achieved.)

6.2.3 NUMERICAL DISPERSION

Similar to the discussion in Chapter 5 deriving spatial numerical dispersion, we again use the simplified (one-dimensional, constant coefficient) partial differential equation with finite difference approximations to the derivatives as a proxy for the compartment approach. Thus, the underlying partial differential equation that we are considering is

$$R \frac{\partial c}{\partial t} = -\bar{U} \frac{\partial c}{\partial x} + E \frac{\partial^2 c}{\partial x^2} \tag{6.22}$$

---

[*] We have not formally derived this nonlinear criterion, but rather made an intuitively analogous extension from the linear case.

[†] It is certainly computationally feasible to monitor the stability criterion during run time and dynamically adjust the time step to ensure stability. Nonetheless, the current version of the GEM uses a fixed, user-supplied time step.

We first derive the numerical dispersion for the FTBS option that involves spatial and temporal numerical dispersion. Following that, we show that the FTCS method involves only temporal numerical dispersion. For the FTBS option, (6.22) becomes (leaving the second derivative term intact for reasons that will be clear later)

$$R\frac{c_i^{t+1} - c_i^t}{\Delta t} = -\overline{U}\frac{c_i^t - c_{i-1}^t}{\Delta x} + E\frac{\partial^2 c}{\partial x^2} \tag{6.23}$$

Expand c_i^{t+1} in a Taylor series about c_i^t and truncate after the second derivative term

$$c_i^{t+1} \approx c_i^t + \Delta t\frac{\partial c}{\partial t} + \frac{\Delta t^2}{2}\frac{\partial^2 c}{\partial t^2} \tag{6.24}$$

Expand c_{i-1}^t in a Taylor series about c_i^t and truncate after the second derivative term

$$c_{i-1}^t \approx c_i^t - \Delta x\frac{\partial c}{\partial x} + \frac{\Delta x^2}{2}\frac{\partial^2 c}{\partial x^2} \tag{6.25}$$

Substituting for c_i^{t+1} from (6.24) and c_{i-1}^t from (6.25) into (6.23), we have

$$R\frac{\left(c_i^t + \Delta t\frac{\partial c}{\partial t} + \frac{\Delta t^2}{2}\frac{\partial^2 c}{\partial t^2}\right) - c_i^t}{\Delta t} = -\overline{U}\frac{\left[c_i^t - \left(c_i^t - \Delta x\frac{\partial c}{\partial x} + \frac{\Delta x^2}{2}\frac{\partial^2 c}{\partial x^2}\right)\right]}{\Delta x} + E\frac{\partial^2 c}{\partial x^2} \tag{6.26}$$

which, after canceling terms and simplifying, is

$$R\frac{\partial c}{\partial t} = -\overline{U}\frac{\partial c}{\partial x} + \left(E + \overline{U}\frac{\Delta x}{2}\right)\frac{\partial^2 c}{\partial x^2} - \frac{R\Delta t}{2}\frac{\partial^2 c}{\partial t^2} \tag{6.27}$$

Now take the $\partial/\partial t$ of both sides of (6.22), i.e.,

$$\frac{\partial\left(R\frac{\partial c}{\partial t}\right)}{\partial t} = \frac{\partial\left(-\overline{U}\frac{\partial c}{\partial x} + E\frac{\partial^2 c}{\partial x^2}\right)}{\partial t}$$

or

$$R\frac{\partial^2 c}{\partial t^2} = -\overline{U}\frac{\partial^2 c}{\partial t\partial x} + E\frac{\partial^3 c}{\partial t\partial x^2} \tag{6.28}$$

Similarly, take the $\partial/\partial x$ of both sides of (6.22) to result in

$$R\frac{\partial^2 c}{\partial x \partial t} = -\bar{U}\frac{\partial^2 c}{\partial x^2} + E\frac{\partial^3 c}{\partial x^3} \qquad (6.29)$$

Substituting from (6.29) into (6.28), we have[*]

$$R\frac{\partial^2 c}{\partial t^2} = -\bar{U}\left(-\frac{\bar{U}}{R}\frac{\partial^2 c}{\partial x^2} + \frac{E}{R}\frac{\partial^3 c}{\partial x^3}\right) + E\frac{\partial^3 c}{\partial t \partial x^2}$$

which can be simplified as

$$R\frac{\partial^2 c}{\partial t^2} = \frac{\bar{U}^2}{R}\frac{\partial^2 c}{\partial x^2} - \frac{\bar{U}E}{R}\frac{\partial^3 c}{\partial x^3} + E\frac{\partial^3 c}{\partial t \partial x^2} \qquad (6.30)$$

Now (finally!), if we substitute from (6.30) into (6.27), we have

$$R\frac{\partial c}{\partial t} = -\bar{U}\frac{\partial c}{\partial x} + \left(E + \bar{U}\frac{\Delta x}{2} - \frac{\Delta t \bar{U}^2}{2R}\right)\frac{\partial^2 c}{\partial x^2} + \text{(remaining terms)} \qquad (6.31)$$

What we see from (6.31) is that we now have two additional terms

$$\bar{U}\frac{\Delta x}{2} - \frac{\Delta t \bar{U}^2}{2R}$$

adding to the true, physical dispersion E. We saw the first term earlier. It is the numerical dispersion introduced by the BS spatial discretization. The second term is new; it is the temporal numerical dispersion introduced by the FT temporal discretization.

It is interesting to note that the temporal numerical dispersion works to *reduce* the spatial numerical dispersion. Thus, by increasing the time step, we can reduce the overall numerical dispersion. However, because the time step is constrained by the stability criterion, it is possible to increase it sufficiently to completely eliminate numerical dispersion only for pure advective systems, i.e., $E_p = 0$.

For FTCS, Equation (6.22) becomes

$$R\frac{c_i^{t+1} - c_i^t}{\Delta t} = -\bar{U}\frac{c_{i+1}^t - c_{i-1}^t}{2\Delta x} + E\frac{\partial^2 c}{\partial x^2} \qquad (6.32)$$

Expand c_i^{t+1} in a Taylor series about c_i^t and truncate after the second derivative term

$$c_{i+1}^t \approx c_i^t + \Delta t\frac{\partial c}{\partial x} + \frac{\Delta x^2}{2}\frac{\partial^2 c}{\partial x^2} \qquad (6.33)$$

---

[*] Note that $\dfrac{\partial^2 c}{\partial x \partial t} \equiv \dfrac{\partial^2 c}{\partial t \partial x}$ is a mathematical identity.

Substituting for c_i^{t+1} from Equation (6.24), c_{i-1}^t from Equation (6.25), and c_{i+1}^t from Equation (6.33) into Equation (6.32), we have

$$R\frac{\left(c_i^t + \Delta t \frac{\partial c}{\partial t} + \frac{\Delta t^2}{2}\frac{\partial^2 c}{\partial t^2}\right) - c_i^t}{\Delta t} =$$

$$-\overline{U}\frac{\left[\left(c_i^t + \Delta x \frac{\partial c}{\partial x} + \frac{\Delta x^2}{2}\frac{\partial^2 c}{\partial x^2}\right) - \left(c_i^t - \Delta x \frac{\partial c}{\partial x} + \frac{\Delta x^2}{2}\frac{\partial^2 c}{\partial x^2}\right)\right]}{2\Delta x} + E\frac{\partial^2 c}{\partial x^2}$$

which, after canceling terms and simplifying, is

$$R\frac{\partial c}{\partial t} = -\overline{U}\frac{\partial c}{\partial x} + E\frac{\partial^2 c}{\partial x^2} - \frac{\Delta t R}{2}\frac{\partial^2 c}{\partial t^2} \tag{6.34}$$

Finally, if we substitute from (6.30) into (6.34), we have

$$R\frac{\partial c}{\partial t} = -\overline{U}\frac{\partial c}{\partial x} + \left(E - \frac{\Delta t \overline{U}^2}{2R}\right)\frac{\partial^2 c}{\partial x^2} + (\text{remaining terms}) \tag{6.35}$$

As expected, we see no spatial numerical dispersion because of the central difference but there is a temporal numerical dispersion. Because the numerical dispersion is negative, it countermands the true physical dispersion resulting in *less* spreading than desired.

As was the case in Chapter 5 and will be throughout the remainder of this chapter, we have been using one-dimensional systems for ease of presentation to describe and quantify error characteristics including numerical dispersion. Numerical dispersion in a single dimension, e.g., x, is generalizable to two- or three-dimensional systems as described in Box 6.2.

**BOX 6.2   NUMERICAL DISPERSION AND E_n
CORRECTIONS IN MULTIPLE DIMENSIONS**

The numerical dispersion in a given dimension induced by the advective flow in a multi-dimensional system is caused by the component of the flow in that dimension. For example, suppose you were modeling a river with a wide section as shown in Figure 6.2 and you want to use a two-dimensional compartment representation at the wide place to enhance resolution, as also shown in Figure 6.2. The true flow directions shown would then be approximated in your compartment grid by flow components in both the x and y directions. Flow in the x direction induces numerical dispersion in that dimension, as we discussed previously. Similarly for flow in the y direction. If you want to correct in both dimensions for numerical dispersion, the dispersion coefficients in both directions must be modified, again as previously described.

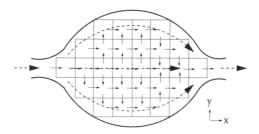

FIGURE 6.2   Two-dimensional compartment grid and flow components.

6.2.4   COMPARISON OF FTBS AND FTCS METHODS

In this section, we compare the BS and CS spatial discretization options when used in the FT dynamic method. First, we compare them assuming no correction for numerical dispersion[*] ($E_m = E_p$). We will refer to these methods as E-unmodified. Then, we correct them for numerical dispersion and assess their differences. We discuss these separately because each strategy has advantages in different applications.

Let's review what we've learned about the FTBS and FTCS methods. For simplicity, we again assume a linear one-dimensional problem with constant coefficients and a single, first-order source or sink term. The following discussion may not be completely generalized to all problems that may be analyzed with the GEM.

Recall from Chapter 5 that a CS method ($\alpha = 0.5$) is subject to a no-wiggle criterion on $L_{i,j}$, i.e., Δx for our simplified problem. That constraint is equally applicable to dynamic problems and is, from Equation (5.41),

$$\Delta x < \frac{E}{(1-\alpha)\overline{U}} \tag{6.36}$$

The stability criterion Equation (6.16) is applicable to both FTBS and FTCS and can be expressed for our simplified problem under consideration as

$$\Delta t < \frac{RV}{\left| Q(1-2\alpha) - 2E' \pm kV\varphi^{T-20} \right|} = \frac{RA\Delta x}{\left| \overline{U}A(1-2\alpha) - 2\dfrac{EA}{\Delta x} \pm kA\Delta x\varphi^{T-20} \right|}$$

$$= \frac{R\Delta x^2}{\left| \overline{U}\Delta x(1-2\alpha) - 2E \pm k\Delta x^2\varphi^{T-20} \right|} \tag{6.37}$$

---

[*] Some perspective may be in order regarding correcting for numerical dispersion. If you *calibrated* your model to existing data to estimate the dispersion coefficients, you will not then need to correct that estimate for numerical dispersion when the model is used for prediction. The estimate that best matches predictions and observations during calibration will have *implicitly* corrected for E_n. It is only when you use an estimate of the *true physical dispersion* (common in chemical risk assessment modeling because there often are no available calibration data) that the correction is needed.

where again | | denotes absolute value. Finally, the numerical dispersion character-istics are

$$\text{FTCS: } E_n = -\frac{\Delta t \overline{U}^2}{2R} \tag{6.38}$$

$$\text{FTBS: } E_n = \frac{\overline{U}\Delta x}{2} - \frac{\Delta t \overline{U}^2}{2R} \tag{6.39}$$

Under our current assumption that $E_m = E_p$, we can see that the advantage of the FTBS method relative to FTCS is that the choice of Δx is unconditionally wiggle-free, i.e., from Equation (6.36)

$$\Delta x < \frac{E_p}{(0)\overline{U}} = \infty$$

Its disadvantage is a smaller allowable time step to ensure stability for a given Δx, i.e., from Equation (6.37)

$$\frac{R\Delta x^2}{\left| -\overline{U}\Delta x - 2E_p \pm k\Delta x^2 \varphi^{T-20} \right|} < \frac{R\Delta x^2}{\left| -2E_p \pm k\Delta x^2 \varphi^{T-20} \right|}$$

Regarding numerical dispersion, which of these options results in less (for the same Δt) depends on the choice of Δx. For values of Δx for which the FTBS spatial E_n is less than twice the temporal E_n, the FTBS will have less total E_n. From Equations (6.38) and (6.39) this can be expressed as

$$\left| \frac{\overline{U}\Delta x}{2} - \frac{\Delta t \overline{U}^2}{2R} \right| < \left| -\frac{\Delta t \overline{U}^2}{2R} \right| \text{ if } \Delta x < 2\overline{U}\Delta t$$

EXAMPLE 6.1

Comparison of E-Unmodified FT Methods

We now illustrate the E-unmodified FTBS and FTCS methods by a series of exam-ples based on numerical simulation of a spill of chemical in a one-dimensional, advective/dispersive aquifer with decay and retardation. Our objective is to illus-trate their relative advantages and compare them to an analytical solution. The example employs linear kinetics and retardation and assumes spatially and tem-porally constant parameters.

We assumes an instantaneous spill of 5 million g benzene to an aquifer that has a cross sectional area of 10,000 m^2. The uniform water content is 0.5 and the aquifer material has a bulk density of 2.5×10^6 g/m³. We assume that benzene has a first-order decay of $k = 0.1$/day and a linear retardation coefficient (K_d) of 1.35×10^{-7} m³/g, as used in a previous example. We assume a constant temperature of 20°C. The aquifer flow rate is 800,000 m³/day and a longitudinal dispersion coefficient of 10,000 m²/day is assumed. The pore water flow velocity is

$$\overline{U} = \frac{Q}{\theta A} = \frac{800,000}{(0.5)(10,000)} = 160 \text{ m/day}$$

The retardation parameter is

$$R = 1 + \frac{(Bd)(Kd)}{\theta} = \frac{(2.5 \times 10^6)(1.35 \times 10^{-7})}{0.5} = 1.68$$

The analytical solution for the case in which the chemical is initially concentrated at distance $x = 0$ is (Schnoor, 1996)

$$c(x,t) = \frac{MR}{2A\sqrt{\pi ERt}} \left(e^{\frac{-\left(x - \overline{U}t/R\right)^2}{4Et/R}} \right) \left(e^{\left(-kt/R\right)} \right) \tag{6.40}$$

where M is the mass of the spill[*] (5×10^6 g) divided by the aquifer cross-sectional flow area, i.e., 10,000 m² times water content. Figure 6.3 shows the progression of the spill demonstrated by the analytical equation at three different times. The progress of the peak concentration downstream over time and its decrease due to longitudinal dispersion and decay can be seen. For purposes of comparison, the figure also shows the concentration profile at 1 day assuming: (1) no decay and (2) no retardation ($K_d = 0$).

We first compare the FTBS and FTCS solutions to the analytical solution using their respective maximum time steps for stability. The first step is to find the maxi-

---

[*] We are simulating *dissolved* chemical in this example. Therefore M is the portion of a *total* spill mass that would immediately become dissolved in the pore water under the standard assumption of instantaneous partitioning. That fraction can be determined as F_d from Equation 2.13, i.e.,

$$F_d = \frac{\theta}{\theta + K_d BD}.$$

For our example problem,

$$F_d = \frac{0.5}{0.5 + (1.35 \times 10^{-7})(2.5 \times 10^6)} = 0.6.$$

Therefore, the total spill mass would be 5 × 10⁶/0.6 = 8.33 × 10⁶ g. It is also noted that F_d and R are related as $F_d = 1/R$ as can be seen by comparing Equations 2.13 and 4.18 when R is rewritten as

$$R = \frac{\theta + K_d BD}{\theta}.$$

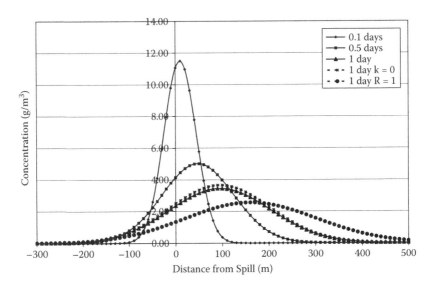

FIGURE 6.3   Analytical solution of spill at three times.

mum compartment length (Δx) that ensures no wiggle for the FTCS option. We then use that length for both FTCS and FTBS options (the E-unmodified FTBS is unconditionally wiggle-free).

From the no-wiggle criterion for the FTCS option, we have a maximum distance between compartment centroids from Equation (6.36) of

$$\Delta x = \frac{E_p}{(1-\alpha)\overline{U}} = \frac{1 \times 10^4}{(1-0.5)160} = 125 \text{ m}$$

We will use 100 m to be conservative. For the FTBS option, the maximum time step from Equation (6.37) is then

$$\Delta t < \frac{R\Delta x^2}{\left|\overline{U}\Delta x(1-2\alpha) - 2E \pm k\Delta x^2\varphi^{T-20}\right|}$$

$$= \frac{1.68(100)^2}{160(100) + 2 \times 10^4 + 0.1(100)^2(1.024^{(20-20)})} = 0.45 \text{ days}$$

and, for the FTCS option, we see a relative advantage in the maximum allowable time step at

$$\Delta t < \frac{R\Delta x^2}{\left|\overline{U}\Delta x(1-2\alpha) - 2E \pm k\Delta x^2\varphi^{T-20}\right|}$$

$$= \frac{1.68(100)^2}{2 \times 10^4 + 0.1(100)^2(1.024^{(20-20)})} = 0.80 \text{ days}$$

We ran the GEM using both the FTCS and FTBS options at time steps of 0.6 days and 0.3 days, respectively. These time steps were chosen to be somewhat conservative

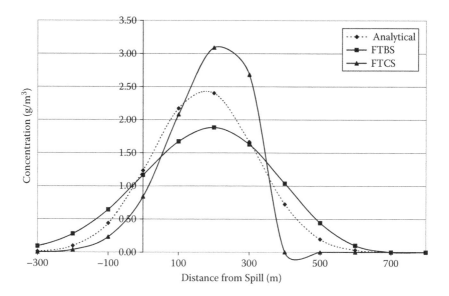

FIGURE 6.4 Numerical dispersion for FT methods.

and to be integer multiples so that results could be compared at the same time after the spill. We used 100 compartments of which 1 and 100 were boundary compartments. The loading is assumed to be in compartment 10 corresponding to $x = 0$. We put the load in compartment 10 instead of 2 (the most upstream interior compartment) so that we could safely assume an upstream boundary condition of 0 in compartment 1. The downstream boundary condition in compartment 100 is also assumed 0.

Alternatively, we could have put the loading in compartment 2 and used dynamic upstream boundary conditions taken from the analytical solution, but our compartment 10 loading approach is simpler. The results are shown in Figure 6.4 at time = 1.8 days after the spill (three time steps for the FTCS and six for the FTBS) and are compared to the analytical solution. The physical dispersion (1×10^4) was used in the model as E_m for both options.

For purposes of the GEM simulations, the spill is represented as an initial condition in the compartment corresponding to $x = 0$ (compartment 10). The initial condition[*] is

$$\frac{5 \times 10^6 \text{g}}{(10,000 \text{ m}^2)(100 \text{ m})(0.5)} = 10 \text{ g/m}^3$$

---

[*] Our use of a discretized numerical model with compartment volumes of 10^6 m³ implies certain early conditions of the spill that we cannot accurately simulate. Our initial condition is 10 g/m³ which represents the spill mass divided by the compartment water volume. Figure 6.3 for the analytical solution shows peak concentrations exceeding 10 g/m³ prior to approximately 0.12 days. (Indeed, at the very instant of the spill before any dilution, transport or decay, the concentration in some very small volume of the aquifer is pure product or 10^6 g/m³.) Therefore, with our compartment volume of 10^6 m³, we cannot reflect these early conditions because our compartment is too large. To do so would require smaller compartment volumes.

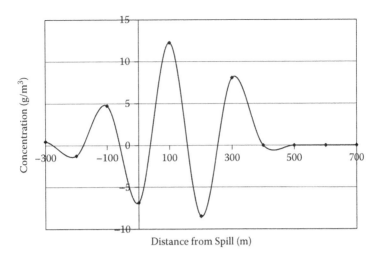

FIGURE 6.5   FTBS with unstable time step.

and represents the initial concentration of *dissolved* benzene in *the pore spaces* of the compartment volume. The numerical dispersion is apparent in Figure 6.4 and it can be seen to decrease the natural dispersion for the FTCS option and increase it for the FTBS option. (Our purpose here is not to show highly accurate simulations. We could do so by decreasing the compartment sizes and time steps, and/or controlling for numerical dispersion. Our purpose is to illustrate and compare limitations of the two FT options.)

To illustrate instability and the relative advantage in stable time steps of the FTCS method versus the FTBS, we ran the FTBS option using the 0.6-day time step that is stable for the FTCS. Results are shown in Figure 6.5 at 0.18 days (three time steps). The instability is obvious.

To illustrate the FTBS method's unconditional no-wiggle advantage, we doubled the compartment lengths to $\Delta x = 200$ m and ran the two methods again. Because the compartment volumes have now doubled, the initial condition used was half the previous value, or 5 g/m³. We used a very stable time step of 0.5 days (under this Δx) for both methods and four time steps. (The maximum time step for stability for the FTBS method under these conditions is 1.2 days[*] and 2.8 days for the FTCS method.) Results are shown in Figure 6.6 along with the analytical solution at 2 days after the spill.

The wiggle in the FTCS solution is apparent; it is not instability (the maximum time step for stability is more than five times the value used). Although the FTBS solution is not particularly accurate because of its numerical dispersion and the now larger Δx discretization, it nonetheless exhibits no wiggle.

We now relax the assumption that $E_m = E_p$ and consider modifying E_m to cancel E_n in both the FTBS and FTCS options. We will refer to these methods as E-modified.

---

[*] Running the FTBS method at the 1.2-day threshold time step for stability nonetheless exhibited some slight instability. Therefore, maximum time steps should be treated somewhat conservatively, just as you treat would the maximum compartment sizes for no wiggles.

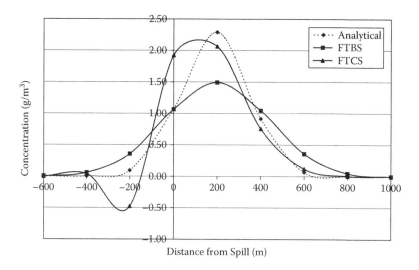

FIGURE 6.6 Wiggle-free FTBS.

Unlike the steady-state CS method (Chapter 5) that has 0 E_n, the FTCS method has a temporal E_n that must be addressed.

For the steady-state case, we saw that correcting the BS option for E_n ($E_m = E_p - E_n$) effectively transforms the BS method into the CS method, including zero numerical dispersion, but also making it subject to wiggle. With that modification, the BS and CS methods were identical (the BS *becomes* the CS method). Let's examine whether that is also the case for the FTBS and FTCS options.

Correcting the FTCS method is straightforward; from Equation (6.38) we *add* the numerical dispersion term

$$E_m = E_p + \frac{\Delta t \overline{U}^2}{2R} \tag{6.41}$$

Because the FTBS E_n has both positive and negative components Equation (6.39), which term will dominate and whether we should add or subtract E_n are not immediately clear. Let's consider what time step yields 0 E_n. We'll denote this as $\Delta t(E_n = 0)$. For any stable time step less than $\Delta t(E_n = 0)$, E_n is positive and we would then *subtract* E_n from E_p to calculate E_m. For any stable time step greater than $\Delta t(E_n = 0)$, E_n is negative and we would *add* E_n to E_p. From Equation (6.39) we can solve for $\Delta t(E_n = 0)$ as

$$\Delta t(E_n = 0) = \frac{R\Delta x}{U}$$

Is $\Delta t(E_n = 0)$ stable, i.e., is

$$\frac{R\Delta x}{\overline{U}} \leq \frac{R\Delta x^2}{\left| -\overline{U}\Delta x - 2E_p \pm k\Delta x^2 \varphi^{T-20} \right|} ?$$

where we have included E_p in the stability criterion because, under our straw man hypothesis of 0 E_n, $E_m = E_p \pm 0$. We can rewrite the right side and express this criterion as: is

$$\frac{R\Delta x}{\overline{U}} \le \frac{R\Delta x}{\overline{U} + \dfrac{2E_p}{\Delta x} \pm |k|\Delta x \varphi^{T-20}} \ ?$$

where the $|k|$ term means use the absolute value of k and $+$ sign if k is a sink (<0) or $-$ sign if a source (>0). For k as a sink ($+|k|$), it can be easily seen that the answer is no and the $\Delta t(E_n = 0)$ time step is unstable. For $k > 0$ (source) and $E_p > 0$, the $\Delta t(E_n = 0)$ time step is also unstable if

$$\frac{2E_p}{\Delta x} - k\Delta x \phi^{T-20} > 0$$

or is

$$k < \frac{2E_p}{\Delta x^2 \phi^{T-20}} \ ?$$

The possibility that a k exceeds the above criterion (resulting in a stable time step of $R\Delta x/\overline{U}$) is highly unlikely in a practical problem. For example, in the present spill problem we have been considering, our E_p is on the order of 10^4 and Δx is on the order of 10^2. Therefore, a k value exceeding this criterion would be on the order of 10^2/day. This is unrealistic* for a first-order growth constant and we are going to dismiss this possibility for practical purposes.

Having established that the maximum time step that is stable is less than $R\Delta x/\overline{U}$, it follows that any time step greater than $R\Delta x/\overline{U}$ is also unstable. Therefore, we know that the E_n from a stable time step will be positive, i.e.,

$$\frac{\overline{U}\Delta x}{2} > \frac{\overline{U}^2 \Delta t_{max}}{2R}$$

because

$$\Delta t_{max} < \frac{R\Delta x}{\overline{U}}.$$

Thus, for FTBS, we *subtract* the numerical dispersion term from the physical dispersion to determine the dispersion coefficient used in the model

$$E_m = E_p - \left(\frac{\overline{U}\Delta x}{2} - \frac{\Delta t \overline{U}^2}{2R} \right) = E_p - \frac{\overline{U}\Delta x}{2} + \frac{\Delta t \overline{U}^2}{2R} \tag{6.42}$$

---

* If this were an analogous continuous compound interest rate problem, for an initial investment of $1, at the end of a day you would have more money than the current U.S. national debt.

Now we can ask are the E-modified FTCS and FTBS options identical for a given Δt and Δx? Yes, because from Equations (6.41) and (6.42) the only term that is different is the BS spatial E_n ($-\overline{U}\Delta x/2$) and we demonstrated in Chapter 5 that modifying E_m for that difference results in equivalence between the BS and CS methods. Here, that same modification will cancel the spatial bias in the FTBS method. The addition of the temporal bias ($U^2\Delta t/2R$) affects *both* methods identically. Therefore, the two are mathematically equivalent.

What about wiggle? Are both methods subject to wiggle? Yes. Similar to the steady-state case, the subtraction of the spatial E_n term in the FTBS E_m parameter transformed the FTBS method into the (E-unmodified) FTCS method that is subject to wiggle. Adding the temporal E_n to the E_m for both methods *increases* the E_m (compared to using $E_m = E_p$). Recall from the development of the criterion in Chapter 5 that wiggle occurs when elements in the off-diagonal of matrix **A** have the same sign (negative for the GEM structure) as the main diagonal elements, and this occurs when (from Equation 5.40)

$$-(1-\alpha_{ij})\left|\overline{Q}_{ij}\right| + E'_{ij} < 0$$

All we are doing by increasing E_m (relative to $E_m = E_p$) is to increase the E' term. It remains a FTCS method, albeit with a larger E' term, just as if we had higher natural dispersion.[*] Depending on its new value relative to the absolute value of flow, it is still possible to result in a negative off-diagonal element value. Therefore, the E-modified FTCS method (and its equivalent, the E-modified FTBS) is subject to wiggle.

Our purpose in this discussion is to eventually demonstrate wiggle-free, stable, numerical dispersion-free solutions using either the E-modified FTBS or FTCS method because they are equivalent. This is a little problematic, however, because our E_m modification introduced a feedback loop between the Δt stability criterion and the Δx no-wiggle criterion, both of which are now functions of E_m, and the E_m equation, which is a function of Δt and Δx. This feedback results in the need for some trial and error by the modeler to find a stable Δt and a wiggle-free Δx.

To make this task easier and eliminate the trial and error, we want to derive a function that expresses a stable Δt as a function of the system parameters E_p and U. We also want to take into account the no-wiggle criterion on Δx. Finally, we want the E_m value used in the model to negate numerical dispersion. Therefore, we are dealing with three simultaneous equations, two of which are inequalities. To keep this analysis tractable, we consider only linear systems [$\underline{f}(\underline{c}) = 0$ and $\mathbf{R} \neq \mathbf{R}(\underline{c})$].

It is difficult to solve simultaneous equations that involve inequalities, so let's introduce a fractional "safety factor" in the stability equation. Let γ be a fractional safety factor for the maximum time step that is stable. We can then write the stability criterion Equation (6.37) for the FTCS method (that is what we are dealing with after the E modification) as the equality

---

[*] It should be clear that our modifications to the FTBS method (1) first transforms it into the equivalent non-E_m-modified FTCS method by subtracting the spatial E_n term (as in Chapter 5), then (2) transforms that result into an E_m-modified FTCS method by adding the temporal E_n term.

$$\Delta t = \frac{\gamma R \Delta x^2}{\left|-2E_m \pm k\Delta x^2 \varphi^{T-20}\right|} \tag{6.43}$$

where $0 < \gamma \le 1$. Similarly, we introduce a safety factor β on the FTCS no-wiggle constraint (6.36) so that the maximum no-wiggle Δx can be written as the equality

$$\Delta x = \beta\left(\frac{2E_m}{\overline{U}}\right) \tag{6.44}$$

where $0 < \beta \le 1$. Substituting (6.44) into (6.43) gives

$$\Delta t = \gamma R \left(\frac{\left(2\beta E_m \middle/ \overline{U}\right)^2}{\left|-2E_m \pm k\left(2\beta E_m \middle/ \overline{U}\right)^2 \varphi^{T-20}\right|}\right) = \frac{2R\beta^2 \gamma E_m}{\overline{U}^2 \pm 2\left|k\right|\beta^2 E_m \varphi^{T-20}} \tag{6.45}$$

where, again, the $\left|k\right|$ term means use the absolute value of k and $+$ sign if k is a sink (<0) or $-$ sign if a source (>0). Now substituting for FTCS E_m (6.41) into (6.45) and grouping terms as coefficients of Δt, we have the quadratic equation

$$\Delta t^2 \left[\pm\frac{\overline{U}^2 \left|k\right|\beta^2 \phi^{T-20}}{R}\right] + \Delta t\left[\overline{U}^2 \pm 2E_p\left|k\right|\beta^2 \phi^{T-20} - \overline{U}^2 \beta^2 \gamma\right] - 2E_p R\beta^2 \gamma = 0 \tag{6.46}$$

Equation (6.46) is the relationship we sought to integrate stability, no-wiggle, and $0 E_n$ for linear systems and applies to both the FTCS and FTBS methods after they are E-modified. To solve this equation for Δt, we recall from high school algebra that the two roots (solutions) of the general quadratic equation $ax^2 + bx + c = 0$ are

$$x = \frac{-b \pm \sqrt{b^2 - 4ac}}{2a}$$

For our application,

$$x = \Delta t$$

$$a = \pm\frac{\overline{U}^2 \left|k\right|\beta^2 \phi^{T-20}}{R}$$

$$b = \overline{U}^2 \pm 2E_p\left|k\right|\beta^2 \phi^{T-20} - \overline{U}^2 \beta^2 \gamma$$

$$c = -2E_p R\beta^2 \gamma$$

We also know that the root of interest to us must be positive, so the correct Δt will be the positive root when k is a sink (denominator is positive) or the negative root when k is a source (denominator is negative).

An alternative equation to Equation (6.46) can be developed by omitting the no-wiggle constraint as shown in Box 6.3. The advantage of the alternative equation is that it allows the user to specify Δx directly, i.e., without using a trial-and-error approach to find that value of β that yields the user's desired Δx. The alternative equation can be derived starting either with the FTBS or FTCS stability criterion and numerical dispersion equation.

BOX 6.3   AN ALTERNATIVE EQUATION TO INTEGRATE STABILITY AND NUMERICAL DISPERSION

Let's start with the FTBS method. For stability, and including our fractional safety factor γ so that we can use an equation instead of an inequality, we have (from Equation 6.37)

$$\Delta t = \frac{\gamma R \Delta x^2}{\overline{U} \Delta x + 2E_m \pm |k| \Delta x^2 \phi^{T-20}} \tag{6.47}$$

where, again, the $|k|$ term means use the absolute value of k and "+" if k is a sink (<0) or "–" if a source (>0).

E_m to control for numerical dispersion is

$$E_m = E_p - \frac{\overline{U} \Delta x}{2} + \frac{\overline{U}^2 \Delta t}{2R} \tag{6.48}$$

where $0 < \gamma \leq 1$. Substituting (6.48) into (6.47) we have

$$\Delta t = \frac{\gamma R \Delta x^2}{\overline{U} \Delta x + 2 \left[E_p - \left(\frac{\overline{U} \Delta x}{2} + \frac{\overline{U}^2 \Delta t}{2R} \right) \right] \pm |k| \Delta x^2 \phi^{T-20}}$$

which can be rewritten as the quadratic equation

$$\frac{\overline{U}^2}{R} \Delta t^2 + \left(2E_p \pm |k| \Delta x^2 \phi^{T-20} \right) \Delta t - \gamma R \Delta x^2 = 0 \tag{6.49}$$

Therefore, given E_p (the true, physical dispersion coefficient) and specifications for Δx and γ, the solution to Equation (6.49) gives us a stable Δt which takes into account the correction for numerical dispersion. The E_m to correct for numerical dispersion is available, after solving (6.49) to get the stable time step, from (6.48). Recall that our E-modified FTBS method is in effect a FTCS method so is subject to wiggle and, because the no-wiggle criterion is not integrated into Equation (6.49), it must be checked for independently from

$$\Delta x < \frac{2E_m}{U}$$

We can similarly apply this technique to the FTCS solution. In that case, the stability equation and numerical dispersion correction equation are, including our safety factor γ,

$$\Delta t = \frac{\gamma R \Delta x^2}{2E_m \pm |k| \Delta x^2 \phi^{T-20}} \tag{6.50}$$

and

$$E_m = E_p + \frac{\overline{U}^2 \Delta t}{2R} \tag{6.51}$$

Substituting (6.51) into (6.50) gives

$$\Delta t = \frac{\gamma R \Delta x^2}{2\left(E_p + \dfrac{\overline{U}^2 \Delta t}{2R}\right) \pm |k| \Delta x^2 \phi^{T-20}} \tag{6.52}$$

which the reader can confirm leads to the same quadratic equation as (6.49), again showing the equivalence of the E-modified FTBS and E-modified FTCS methods.

EXAMPLE 6.2

Demonstration of Zero Numerical Dispersion from E-Modified FT Methods

Returning to our objective of showing stable, wiggle-free, numerical dispersion-free results, we ran the aquifer chemical spill problem for both the E-modified FTCS and FTBS methods using a time step provided by Equation (6.49) from Box 6.3. For consistency with the previous (E-unmodified) results, we specified $\Delta x = 100$ m. (You could also, by trial-and-error use of Equation (6.46), find that $\beta = 0.642$ results in $\Delta x = 100$.) We also specified $\gamma = 0.5$ to be conservative with respect to stability. The resulting time step is $\Delta t = 0.324$ days and was used for both methods. For the FTCS simulation,

$$E_m = 1.0 \times 10^4 + \frac{(160)^2(0.324)}{2(1.68)} = 1.25 \times 10^4$$

For the FTBS simulation, E_m is

$$E_m = 1.0 \times 10^4 - \frac{(160)(100)}{2} + \frac{(160)^2(0.324)}{2(1.68)} = 4.47 \times 10^3$$

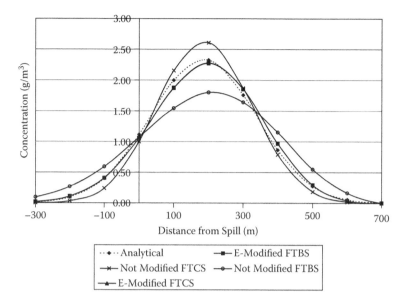

FIGURE 6.7   Zero numerical dispersion from E-modified FTCS and FTBS.

We ran six time steps and the results at 1.944 days are shown in Figure 6.7. Note that the two solutions are identical and compare very well with the analytical solution. Numerical dispersion is not an issue; the relatively minor inaccuracies are attributable to discretization errors at this time step and Δx. For comparison purposes, we also ran the E-unmodified FTBS and FTCS methods at the same time step and those results are also shown in the figure. The numerical dispersion is again apparent.

The spill scenario evaluated in Example 6.2 is an interesting problem and one that we will continue with throughout the remainder of this chapter. It is also an inherently dynamic problem. Steady-state conditions make no sense for the implications of a spill. However, we will take a brief detour from our spill to consider in Box 6.4 a different type of dynamic problem—the time to steady state for a contamination scenario with constant loadings and parameters. Our purpose here is to illustrate the time to steady state and show GEM dynamic solutions for a different type of problem.

BOX 6.4   TIME-TO-STEADY-STATE EXAMPLE

This example uses the same aquifer considered in the spill example and the same GEM configuration of that aquifer. Instead of a spill-type loading, here we consider the scenario of a constant temporal loading of chemical. Rather than introduce the loading as the initial condition as we did for the spill example, the loading here enters the solution as an upgradient boundary condition, i.e., the constant concentration associated with compartment 1.

The analytical solution where there is a constant input of chemical through a boundary condition at distance $x = 0$ is (Schnoor, 1996)

$$c(x,t) = \frac{c_0}{2} e^{\frac{(\overline{U}-v)x}{2E}} erfc\left[\frac{Rx - vt}{2\sqrt{ERt}}\right] + \frac{c_0}{2} e^{\frac{(\overline{U}+v)x}{2E}} erfc\left[\frac{Rx + vt}{2\sqrt{ERt}}\right]\quad (6.53.a)$$

where

$$c(0,t) = c_0 \text{ for } t > 0 \text{ is the boundary condition} \quad (6.53.b)$$

$$v = \overline{U}\left(1 + \frac{4kE}{\overline{U}^2}\right)^{1/2} \quad (6.53.c)$$

and *erfc* denotes the mathematical error function. Values of the error function at various values are available in mathematical tables and commonly integrally coded in spreadsheets and other software.

Figure 6.8 shows the results of running the analytical model (6.53) for three different times and compares the GEM solution to the analytical results at the same times. For the GEM, the FTCS method was used with the same parameters as were used for Example 6.2, i.e., $\Delta t = 0.324$ days for stability and $E_m = 1.25 \times 10^4$ m²/day to control numerical dispersion. However, the initial conditions were 0 everywhere and we used 10 g/m3 as the boundary condition in the GEM's compartment 1 and c_0 in the analytical model. For the first run, six time steps made a total time of 1.944 days. The second run was 60 time steps to obtain 19.44 days. The third run used the GEM's steady-state mode. The analytical model was evaluated at the same total times for runs 1 and 2, and at time 1×10^6 days for the steady-state scenario. The agreement between GEM predictions and the analytical results were very close. This sort of example also illustrates one method of estimating the time to steady state for such problems. As can be seen from Figure 6.8, the 19.44 day simulation is clearly approaching steady state.

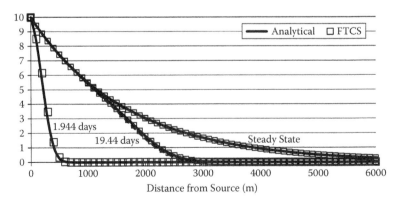

FIGURE 6.8   Time-to-steady-state comparisons from E-modified FTCS.

The E modification produces the very desirable effect of eliminating numerical dispersion and also provides a mechanism whereby the modeler can greatly increase the stable time step and/or spatial disaggregation relative to the non-E-modified methods if desired. The relationship between the stable time step and wiggle-free length relationship given by Equation (6.46) is positive (as Δx increases Δt increases and vice versa). This arises because E_m is now a function of Δx and Δt unlike in the non-E-modified scenario where it is simply $E_m = E_p$. When $E_m = E_p$, for the FTCS method, one is "stuck" with a wiggle-free Δx determined entirely by the E_p fixed parameter. When $E_m = E_p$ for both the FTCS and FTBS methods, one is also stuck with a stable Δt that is also determined entirely by fixed system parameters. With the E_m-modification, it is possible to increase Δx and/or Δt considerably, at least up to the limits of $0 < \gamma$ and $\beta \leq 1$ and $\beta^2 \gamma < 1$.

To illustrate this feature, Table 6.1 is a matrix showing the stable solution Δt of Equation (6.46) and the corresponding wiggle-free Δx from Equation 6.44 for a range of values of γ and β for our aquifer chemical spill example. Also shown are the E_m values that would be used in E-modified FTCS and FTBS simulations. For each γ and β combination in the table, four values are presented. The upper left value (e.g., 1.8E-01 for $\gamma = \beta = 0.5$) is the time step. The upper right value (e.g., 7.1E+01 for $\gamma = \beta = 0.5$) is the Δx. The lower left value is the E_m to be used for the FTCS method while the lower right value is E_m to be used for FTBS. What we see is that both Δt and Δx increase as γ and/or β increase, i.e., become less conservative. If you want a relatively high spatial resolution and a relatively low temporal resolution you would use a smaller β and larger γ and vice versa. To achieve a low spatial and temporal resolution, you would use a larger β and γ. For example, at the extreme (β and $\gamma =$ 0.999), the wiggle-free or stable[*] Δx and Δt are impressive 510 m and 4.1 day results, respectively (as compared to 125 m from the E-unmodified FTCS and 0.45 days from the E-unmodified FTBS).

Warning to the reader: The following discussion revisits the pure advection problem[†] under dynamic conditions. Unfortunately, it is somewhat detailed and includes several examples. If you are a casual reader or not particularly interested in pure advection, you may want to skip it. By way of a summary for those electing to jump forward, recall from Chapter 5 that the only viable approach to the pure advection problem under steady-state conditions was the BS method. The CS method (or equivalently the E-modified BS method) was unconditionally wiggly and also suffered from the downstream boundary condition problem. Neither problem

---

[*] With β and $\gamma = 0.999$, however, you should expect to see *some* (perhaps considerable) wiggle and instability. Unfortunately, the stability and no-wiggle criteria are not bright lines where β and $\gamma = 0.999$ absolutely do not violate them and β and $\gamma = 1.001$ do violate them.

[†] We are not trying to be overly obsessed with the pure advection problem. Many environmental problems do not consider dispersion because it is an appropriate physical assumption or because the assumption is conservative for the analysis objectives. In addition, the pure advection problem is an interesting and challenging test of the GEM methods that are in fact based on solving the advection/dispersion problem. For interested readers, the "pure advection" problem is discussed in the applied mathematics literature (e.g., Hoffman, 1992) often under the title of "convection" problem. When only convective (our advective) transport is considered, the underlying partial differential equation is termed homogeneous. When sources and sinks (called forcing functions) are included, the partial differential equation is termed non-homogeneous.

TABLE 6.1
Variation of Stable and Wiggle-Free Δx and Δt with γ and β

		β (affects wiggle) 0.5		β (affects wiggle) 0.6		β (affects wiggle) 0.7	
γ (affects stability)	0.5	1.8E–01	7.1E+01	2.8E–01	9.1E+01	4.0E–01	1.1E+02
		1.1E+04	5.7E+03	1.2E+04	4.8E+03	1.3E+04	3.9E+03
	0.6	2.3E–01	7.3E+01	3.5E–01	9.5E+01	5.1E–01	1.2E+02
		1.2E+04	5.9E+03	1.3E+04	5.1E+03	1.4E+04	4.2E+03
	0.7	2.7E–01	7.5E+01	4.2E–01	9.9E+01	6.3E–01	1.3E+02
		1.2E+04	6.0E+03	1.3E+04	5.3E+03	1.5E+04	4.4E+03
	0.8	3.2E–01	7.8E+01	5.0E–01	1.0E+02	7.7E–01	1.4E+02
		1.2E+04	6.2E+03	1.4E+04	5.5E+03	1.6E+04	4.8E+03
	0.9	3.7E–01	8.0E+01	5.9E–01	1.1E+02	9.3E–01	1.5E+02
		1.3E+04	6.4E+03	1.5E+04	5.8E+03	1.7E+04	5.1E+03
	0.999	4.2E–01	8.3E+01	6.9E–01	1.1E+02	1.1E+00	1.6E+02
		1.3E+04	6.6E+03	1.5E+04	6.1E+03	1.8E+04	5.5E+03

		β (affects wiggle) 0.8		β (affects wiggle) 0.9		β (affects wiggle) 0.999	
γ (affects stability)	0.5	5.6E–01	1.4E+02	7.6E–01	1.8E+02	1.0E+00	2.2E+02
		1.4E+04	2.9E+03	1.6E+04	1.6E+03	1.8E+04	1.8E+01
	0.6	7.3E–01	1.6E+02	1.0E+00	2.0E+02	1.4E+00	2.6E+02
		1.6E+04	3.1E+03	1.8E+04	1.8E+03	2.1E+04	2.1E+01
	0.7	9.2E–01	1.7E+02	1.3E+00	2.3E+02	1.9E+00	3.0E+02
		1.7E+04	3.4E+03	2.0E+04	2.0E+03	2.4E+04	2.4E+01
	0.8	1.2E+00	1.9E+02	1.7E+00	2.6E+02	2.5E+00	3.6E+02
		1.9E+04	3.8E+03	2.3E+04	2.3E+03	2.9E+04	2.9E+01
	0.9	1.4E+00	2.1E+02	2.2E+00	3.0E+02	3.2E+00	4.3E+02
		2.1E+04	4.2E+03	2.7E+04	2.7E+03	3.4E+04	3.4E+01
	0.999	1.8E+00	2.3E+02	2.7E+00	3.5E+02	4.1E+00	5.1E+02
		2.3E+04	4.7E+03	3.1E+04	3.1E+03	4.1E+04	4.1E+01

affects the E-unmodified BS and we concluded that the E-unmodified BS was the clearly preferred approach. A similar conclusion is the case for the FTBS and FTCS dynamic methods as we show in the following.

EXAMPLE 6.3

Pure Advection

Our interest in this example is to examine the effect on the no-wiggle Δx for an E-modified FT method when E_p is decreasing to 0 (approaching the pure advection scenario).

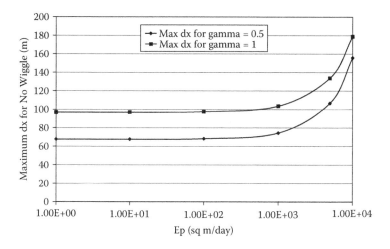

FIGURE 6.9 Maximum Δx as $f(E_p)$.

We used Equation (6.49) from Box 6.3 for our aquifer chemical spill problem to illustrate (Figure 6.9) how the no-wiggle, maximum Δx decreases as E_p decreases. As before, our desired Δx is 100 m, $k = -0.1$, and $R = 1.68$. We used $\gamma = 0.5$ to ensure stability, as previously. We also used $\gamma = 1.0$ to analyze the extreme. At $E_p = 1 \times 10^4$ (as used previously), our desired $\Delta x = 100$ is well within the no-wiggle limits.

However, the no-wiggle limit begins to drop sharply as we decrease E_p to 1×10^3, reaching an asymptote of approximately 68 m (for $\gamma = 0.5$) as E_p drops below 1×10^3. Even using the non-conservative $\gamma = 1.0$, the asymptote is slightly below our desired $\Delta x = 100$ m. In other words, if we have a pure advection problem or are even approaching that condition and want $\Delta x = 100$ m, we cannot satisfy the no-wiggle criterion for our example.

Suppose we tried to vary our desired Δx to find a smaller value than 100 m that will meet the no-wiggle limit, and we are even willing to use $\gamma = 1.0$. Figure 6.10 shows the relationship between our desired Δx and the maximum, no-wiggle Δx

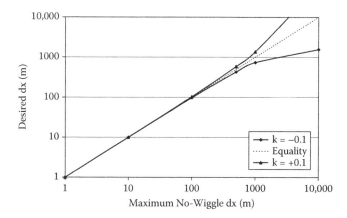

FIGURE 6.10 Relationship between desired and maximum Δx.

for $k = -0.1$. Also shown is a 45-degree line that illustrates that our goal is always just out of reach. For more negative k values, the situation gets even worse. For desired Δx values greater than 100 m (also shown in Figure 6.10), the situation also gets worse.

We can see this no-wiggle problem more generally. From our quadratic Equation (6.46) used to find that time step eliminating numerical dispersion while simultaneously satisfying user-specified safety factors on stability and no-wiggle, we can solve for the time step explicitly when $E_p = 0$ as

$$\Delta t = \frac{R\left(\gamma - \frac{1}{\beta^2}\right)}{\pm |k| \varphi^{T-20}} \tag{6.54}$$

where, again, the $|k|$ term means use the absolute value of k and a + symbol is used if k is a sink (<0) and a – sign if a source (>0).

Just as Equation (6.46) yielded a stable Δt as a function of γ (affecting stability) and β (affecting wiggle) for the general advective/dispersive problem, Equation (6.54) gives it for the pure advection problem. For $k < 0$ (decaying chemical spill problem), Δt is positive only when

$$\left(\gamma - \frac{1}{\beta^2}\right) > 0$$

or equivalently when

$$\beta > \sqrt{\frac{1}{\gamma}}$$

Because, to ensure stability and no-wiggle, $0 < \gamma$ and $\beta \leq 1$, a moment's reflection will convince you that no combination of γ and β meets this condition. The only stable Δt is 0 for $\gamma = \beta = 1$, which is not very helpful.

For $k > 0$ (source), Δt is positive when

$$\left(\gamma - \frac{1}{\beta^2}\right) < 0$$

or when

$$\beta < \sqrt{\frac{1}{\gamma}}$$

which would not seem to be an issue. However, as shown in Figure 6.10 for $k = +0.1$, while our desired Δx is less than the no-wiggle Δx throughout much of the range shown, it is not very much less, especially for reasonably small values of Δx that would be used in practice. (At desired Δx of 100, the no-wiggle Δx is 103 m.) For more conservative values of γ (than 1.0), this difference gets even smaller. For example, at $\gamma = 0.9$, the no-wiggle Δx is 98 m. Only as you move toward large desired Δx values does significant $\beta < 1$ breathing room become available, but that leads to a different problem.

Previously we cited as advantages of the E modification that (1) we were not stuck with a time step and/or spatial disaggregation solely determined by fixed system parameters, and (2) the modified E value can greatly increase Δx and Δt relative to using $E_m = E_p$. (Again, because the relationship between the stable time step/wiggle-free length relationship given by Equation (6.46) is positive; as Δx increases Δt increases and vice versa.)

The coupling of Δx and Δt now becomes a disadvantage because, at the large Δx values needed to achieve any no-wiggle breathing room for the pure advection problem, the associated Δt is now extremely large and leads to the fast kinetics discretization error problem cited in Chapter 4. For example, at the 1000 m Δx in Figure 6.10 that provides some no-wiggle breathing room, the associated Δt for the E-modified FTCS method is 14.3 days ($\gamma = 1$ and $E_m = 1.09 \times 10^5$). At that very large time step, major discretization error would result, even for a relatively slow rate constant. Finally, the downstream boundary condition problem discourages use of the E-modified FTCS and FTBS methods for the pure advection problem. In short, the E-modified FT methods are ill-suited to the dynamic pure advection problem.

Recalling our relative success with the E-unmodified BS method for steady-state, pure advection problems, let's now examine the E-unmodified FTBS method. The numerical dispersion is given by

$$E_n = \frac{\overline{U}\Delta x}{2} - \frac{\overline{U}^2 \Delta t}{2R}$$

Because these two terms tend to offset each other, we saw earlier that a time step that will eliminate E_n can be found by setting E_n to 0 in the above equation and solving for the time step as

$$\Delta t = \frac{R\Delta x}{\overline{U}}$$

For $k = 0$, we see that the $0\ E_n$ time step

$$\Delta t = \frac{R\Delta x}{\overline{U}}$$

is *identically* the maximum time step that is stable. (Compare with Equation (6.18).)

EXAMPLE 6.4

E-Unmodified FTBS for Pure Advection and Conservative Chemical

To illustrate this result, consider our aquifer spill problem but with no chemical decay ($k = 0$). All other parameters are as previously used. The analytical solution for this problem is simple. The initial concentration simply translates downstream

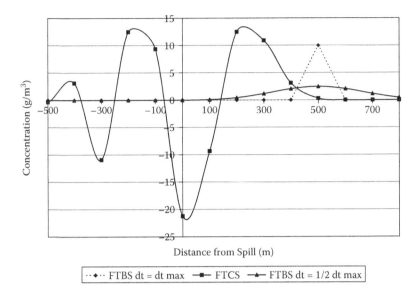

Distance from Spill (m)

··•·· FTBS dt = dt max   ■ FTCS   ▲ FTBS dt = 1/2 dt max

FIGURE 6.11   E-unmodified FTBS for pure advection and conservative chemical.

at a rate equal to the pore water velocity divided by the retardation factor. A useful way to think of the retardation factor is (Schnoor, 1996):

$$R = \frac{\overline{U}}{\text{chemical velocity}}$$

For example, after 5 days the pore water velocity would have covered 800 m while the chemical would have moved 800/1.68 = 476 m. The assumption of no dispersion prevents longitudinal mixing so the initial spike remains a spike. This is called a "plug flow" system.

We ran the GEM under the $k = 0$ assumption using the E-unmodified FTBS method and its maximum time step for stability,

$$\Delta t = \frac{R\Delta x}{\overline{U}} = \frac{1.68(100)}{160} = 1.0465 \text{ days.}$$

$E_m = E_p = 0$ was used. Figure 6.11 shows the concentration profile after five time steps. As can be seen, the initial condition has translated downstream some 476 m after five time steps. (The spike made it into compartment 15, which is between 400 and 500 m from the spill compartment—compartment 10.) Also shown is a run using the E-unmodified FTCS method[*] using the same time step, compartment sizes, and $E_m = 0$ after five time steps. As expected, it exhibits wiggle because $E_m = 0$. Finally, a run using the E-unmodified FTBS with the time step equal to half

---

[*] In this spill application, the downstream boundary condition required for a CS method is known (0), at least until the dispersion wave begins to approach the downstream boundary compartment.

the maximum time step above is shown after 10 time steps. The point here is that this is a stable time step but because it is less than the maximum stable time step it induces numerical dispersion, which is clearly seen.

Recall in the discussion of the pure advection problem under steady-state conditions in Chapter 5 that, despite the Taylor series analysis indicating the presence of numerical dispersion for the BS method when $E_p = 0$, we concluded that there is no numerical dispersion. For the dynamic case, in the preceding Example 6.4, we just demonstrated zero numerical dispersion when the maximum stable time step is used. This was for a conservative chemical and the zero numerical dispersion result is predictable by the E_n equation

$$E_n = \frac{\overline{U}\Delta x}{2} - \frac{\overline{U}^2 \Delta t}{2R}$$

discussed above. There is no inconsistency of that result with the Taylor series analysis for the conservative chemical case.

We now make an argument that using the maximum stable time step and the E-unmodified FTBS method also eliminates numerical dispersion for non-conservative ($k \neq 0$) chemical assumptions, which is inconsistent with the E_n predictive equation. Again, we see no reason why the Taylor series analysis is not applicable to the $E_p = 0$ case, but apparently it is not. The argument is that the maximum stable time step is the unique step that moves all of the chemical mass out of a compartment i and into an adjacent compartment j in a single time step (just as happens in reality in a plug flow system with an arbitrarily small compartment volume).

Any smaller time step would leave some of the mass behind in compartment i, and this residual mass constitutes the numerical dispersion. Any larger time step would be unstable. To illustrate, the FTBS mass balance equation for our one-dimensional aquifer spill problem with constant coefficients, 20°C temperature, and $E_p = 0$ can be written as for time t as

$$RV\frac{c_i^{t+1} - c_i^t}{\Delta t} = Qc_{i-1}^t - Qc_i^t \pm |k|Vc_i^t$$

Let's assume the spill is in compartment i at time t, and we seek a single time step that will completely advect the mass downstream into the next compartment. Under this assumption, we know that $c_i^{t+1} = c_{i-1}^t = 0$ because the spill will have left compartment i after time t and it was either never in compartment $i - 1$ at time t or it already left. We can then write the above mass balance equation as

$$\frac{RV}{\Delta t}c_i^t = c_i^t(Q \pm |k|V)$$

which can be solved for our desired time step as

$$\Delta t = \frac{RV}{Q \pm |k| V}$$

The reader can confirm that this is identically the maximum stable time step.

EXAMPLE 6.5

E-Unmodified FTBS for Pure Advection and Non-Conservative Chemical

To illustrate these results, we ran the E-unmodified FTBS method on the plug flow aquifer spill problem using k as a source and a sink. For $k = 0.1$, the maximum time step is

$$\Delta t = \frac{R \Delta x}{\overline{U} \pm |k| \Delta x \varphi^{T-20}} = \frac{1.68(100)}{160 - 0.1(100)(1)} = 1.11667$$

while for $k = -0.1$, it is

$$\Delta t = \frac{R \Delta x}{\overline{U} \pm |k| \Delta x \varphi^{T-20}} = \frac{1.68(100)}{160 + 0.1(100)(1)} = 0.98529$$

Seven maximum time steps were made for each run, and these results are shown in Figure 6.12 at elapsed total times of 7.816 and 6.897 days, respectively. $E_m = E_p = 0$. Also shown in the figure are the analytical results obtained from

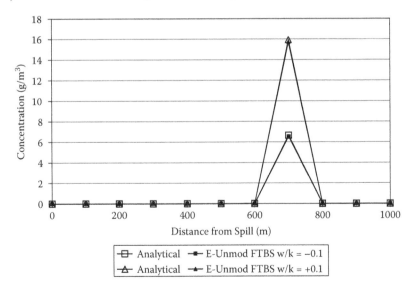

FIGURE 6.12   E-unmodified FTBS for pure advection and non-conservative chemical.

$$c(t) = c(0)e^{\left(\pm kt/R\right)}$$

where $c(0) = 10$ and the retardation factor R modifies the pore water time of travel to the chemical time of travel. This equation is for a batch reactor, not a plug flow system. We are simply assuming that the batch reactor is advecting downstream in a plug flow manner. As can be seen, the FTBS results are excellent compared with the exact solutions. There is no evidence of numerical dispersion and the peak concentrations are nearly exactly matched.

In summary, just as we found for the pure advection steady state problem, the E-unmodified FTBS method using the maximum time step that is stable appears well suited for the dynamic problem. Again, however, the use of the maximum time step to avoid numerical dispersion will cause fast kinetics problems when the spatial disaggregation is coarse. For example, if $k = +0.5/$day for our spill example and we continue to use $\Delta x = 100$ m, our maximum time step is

$$\Delta t = \frac{R\Delta x}{U \pm |k| \Delta x \varphi^{T-20}} = \frac{1.68(100)}{160 - 0.5(100)(1)} = 1.5227$$

and that relatively large time step will lead to the discretization errors shown in Figure 6.13 after four time steps at 6.092 days, as compared to the analytical solution. Not only has the exact concentration

$$\left(10e^{\left(0.5*6.092/1.68\right)} = 61.3\right)$$

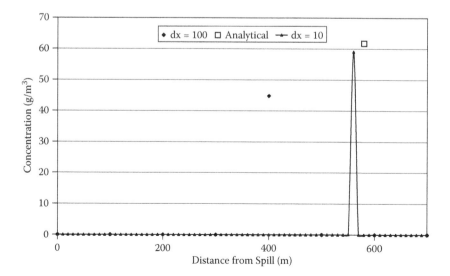

FIGURE 6.13 Fast kinetics problem.

been underestimated, but its location has been underestimated. After 6.092 days, the plug (again, subject to retardation) advanced from $x = 0$ m to

$$x = \frac{(160)(6.092)}{1.68} = 580 \text{ m},$$

whereas the GEM solution has the plug with concentration 44.8 g/m^3 in compartment 14 ($x = 350$ to $x = 450$). The cure for this problem is to use a smaller Δx. For example, at $\Delta x = 10$ m, the maximum time step is 0.108 days and the results of using this smaller spatial and temporal discretization are also shown in Figure 6.13 after 56 time steps at time 6.051 days. The solution now reveals only minor discretization error.

6.3   EXPLICIT CENTERED TIME (MacCORMACK) METHOD

6.3.1   METHODOLOGY

The MacCormack method (1969) is the second explicit approach we will consider. It has the same advantages of the Euler method for avoiding solutions of simultaneous equations at each time step and robustness to $\underline{f}(\underline{c})$-type nonlinear problems (but not $\mathbf{R}(\underline{c})$–type nonlinearities). Its advantages relative to the Euler method are somewhat enhanced stability conditions and greatly reduced (but not zero) numerical dispersion. The MacCormack method attains these advantages as a result of a two-step, predictor–corrector approach that approximates a temporal centered difference.

It is helpful in understanding the MacCormack approach to consider again the general system of equations in (4.22) and repeated below:

$$\frac{d(\mathbf{R}(\underline{c})\mathbf{V}\underline{c})}{dt} = \mathbf{A}\underline{c} + \underline{f}(\underline{c}) + \underline{b} \qquad \text{(4.22 repeated)}$$

The right side of this equation can be thought of as the slope of a curve that plots $\mathbf{R}(\underline{c})\mathbf{V}\underline{c}$ versus time. In other words, consider the simple equation $c(t) = at$ where t is time and a is some arbitrary constant. A plot of c versus time is a line with slope a. That is, the slope of this line is calculated as $dc/dt = a$. Just as the right side of this simple equation is the slope of the function $c(t) = at$, $\mathbf{A}\underline{c} + \underline{f}(\underline{c}) + \underline{b}$ is the slope of $\mathbf{R}(\underline{c})\mathbf{V}\underline{c}$ versus time. With this perspective, we first consider the predictor step of the MacCormack method.

The predictor step is identical to the Euler method. Recall that the forward time Euler method is based on the following approximation to the general system of equations:

$$\frac{(\mathbf{R}(\underline{c})\mathbf{V}\underline{c})^{t+1} - (\mathbf{R}(\underline{c})\mathbf{V}\underline{c})^{t}}{\Delta t} = (\mathbf{A}\underline{c} + \underline{f}(\underline{c}) + \underline{b})^{t} \qquad \text{(6.7 repeated)}$$

As can be seen from Equation (6.7), the derivative or slope within the time interval t to $t + 1$ is assumed to be approximated by $\mathbf{A}\underline{c} + \underline{f}(\underline{c}) + \underline{b}$ *evaluated at time t.*

Presumably, an equally valid estimate of this slope within the time interval would be the same term evaluated at the end of the interval $t + 1$. Indeed, the two implicit methods considered later use this slope. Furthermore, if these two estimates of the true slope are equally valid (albeit both involve some error due to discretization), then a reasonable compromise may be to use their average. (This temporal averaging is the reason that the temporal numerical dispersion is greatly reduced relative to methods that use one or the other endpoint.) Thus, with this approach, we can modify Equation (6.7) as

$$(\mathbf{R}(\underline{c})\mathbf{V}\underline{c})^{t+1} = (\mathbf{R}(\underline{c})\mathbf{V}\underline{c})^{t} + \Delta t \frac{[(\mathbf{A}\underline{c} + \underline{f}(\underline{c}) + \mathbf{b})^{t} + (\mathbf{A}\underline{c} + \underline{f}(\underline{c}) + \mathbf{b})^{t+1}]}{2} \quad (6.55)$$

Recall that with respect to this equation, our solution has been advanced to time t (all concentrations up to and including time t are known), while concentrations at $t + 1$ (and later) are as yet unknown. Therefore, Equation (6.55) cannot be solved explicitly because of the concentrations at time $t + 1$ on the right side, i.e., the slope at $t + 1$. To get around this difficulty,[*] the first stage of the MacCormack method uses the (Euler-like) explicit predictions as an *estimate* of $\underline{c}$ at time $t + 1$. These estimates are then substituted into the $(\mathbf{A}\underline{c} + \underline{f}(\underline{c}) + \mathbf{b})^{t+1}$ slope term, so that it is evaluated (albeit with some prediction error) at time $t + 1$. With this interesting approach, we can then rewrite Equation (6.55) as

$$(\mathbf{R}(\underline{c})\mathbf{V}\underline{c})^{t+1} = (\mathbf{R}(\underline{c})\mathbf{V}\underline{c})^{t} + \Delta t \frac{[(\mathbf{A}\underline{c} + \underline{f}(\underline{c}) + \mathbf{b})^{t} + (\mathbf{A}\hat{\underline{c}} + \underline{f}(\hat{\underline{c}}) + \mathbf{b})^{t+1}]}{2} \quad (6.56)$$

where $\hat{\underline{c}}$ represents the estimated concentrations from the first (predictor) step.

In a further attempt to reduce numerical dispersion, the MacCormack method uses different values of the spatial differencing parameter (advective transport weighting parameter α) for the predictor step versus the corrector step to mitigate spatial numerical dispersion. The GEM uses $\alpha = 0$ (forward spatial difference) for the predictor slope estimate and $\alpha = 1$ (backward spatial difference) for the corrector slope estimate, as indicated in Equation (6.56). A time and space stencil of the MacCormack method is shown in Figure 6.14.

This could be reversed, and some implementations of MacCormack's method for highly dispersive systems use centered differences (Hoffman, 1992). As we have seen, while methods that use a biased spatial difference ($\alpha = 0$ or $\alpha = 1$) introduce spatial numerical dispersion as a result, the MacCormack approach of alternating between these endpoints effectively reduces spatial numerical dispersion. This is in addition to the reduction in temporal numerical dispersion achieved by the averaging of the two endpoint slope terms.

---

[*] We could solve this implicitly as we do for the two subsequent methods. Note that this would not be equivalent to either of the two subsequent implicit methods discussed. It would be an interesting exercise to carry out the methods and compare accuracy characteristics.

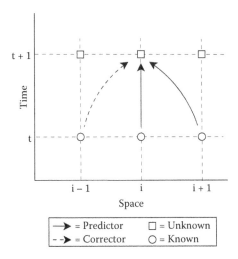

FIGURE 6.14   Time and space stencil of MacCormack method.

For problems not involving $\mathbf{R(\underline{c})}$-type nonlinearities, we can now write Equation (6.56) in explicit form as

$$\underline{c}^{t+1} = \left((\mathbf{RV})^{-1}\right)^{t+1}\left((\mathbf{RV}\underline{c})^{t}\right)$$
$$+ \frac{\Delta t}{2}\left\{\left[\mathbf{A}(\alpha = 0)\underline{c} + \underline{f}(\underline{c}) + \underline{b}\right]^{t} + \left[\mathbf{A}(\alpha = 1)\hat{\underline{c}} + \underline{f}(\hat{\underline{c}}) + \underline{b}\right]^{t+1}\right\}$$

(6.57)

For systems involving $\mathbf{R(\underline{c})}$-type nonlinearities, we must still solve systems of nonlinear equations. We don't see that MacCormack's method in a nonlinear implementation offers any advantages over other methods presented and indeed has the distinct disadvantage that nonlinear systems must be solved twice—once for the predictor step and again for the corrector. Therefore, the GEM does not include an $\mathbf{R(\underline{c})}$-type nonlinear option for the MacCormack method.

6.3.2   Stability, Numerical Dispersion, and Wiggle

While it appears relatively simple, the MacCormack method is nonetheless "… too complicated to yield simple exact stability criteria" (Hoffman, 1992) for either linear or nonlinear systems. Accordingly, we will simply state here that the method is more stable than the forward time, Euler but is not unconditionally stable.

As discussed above, the MacCormack method is designed to reduce both spatial and temporal numerical dispersion by means of its approximate centering in space and time. However, the method is not fully centered and we will simply state again that it enjoys less numerical dispersion than the forward time Euler method but some dispersion does occur.

The MacCormack method is subject to wiggle, because of its $\alpha = 0$ for the predictor slope estimate. In fact, from the no-wiggle criterion,

$$L_{ij} < \frac{E_{ij}}{(1-\alpha_{ij})U_{ij}}$$ (5.41 repeated)

We can see that $\alpha = 0$ requires a compartment length half that of a CS method in which $\alpha = 0.5$. This disadvantage is mitigated but not eliminated by the $\alpha = 1$ (not subject to wiggle) corrector slope estimate and subsequent averaging of results.

EXAMPLE 6.6

MacCormack versus E-Unmodified FT Methods

To illustrate the enhanced stability and reduced numerical dispersion character-istics of the MacCormack method as compared to the E-unmodified FTBS and FTCS methods, we ran the aquifer spill problem ($E_p = 1 \times 10^4$, $\overline{U} = 160$, $R = 1.68$, and $k = -0.1$) using the MacCormack method with a time step of 0.8 days and three time steps, and again used $E_m = E_p$. $\Delta x = 100$ m. The 0.8-day time step is equal to the maximum stable time step for the FTCS method, which has the larger stable time step of the two E-unmodified FT methods. (Previously, we used a time step of 0.6 days for the FTCS method, to be conservative.) The result is shown in Figure 6.15 and is compared to the analytical solution and the concentration profiles from the E-unmodified FTBS and FTCS solutions at the same (0.8-day) time step.

Note that although some wiggle[*] or instability is evident in the MacCormack solution, it is greatly diminished relative to the instabilities exhibited by the FT methods. The oscillating errors in the FT methods are instabilities, not wiggles. The FTBS method is unconditionally wiggle-free and the FTCS method satisfies

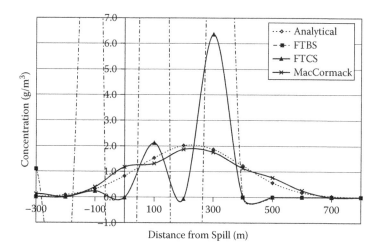

FIGURE 6.15 Enhanced stability and diminished E_n from MacCormack method.

---

[*] $\Delta x = 100$ m length violates the no-wiggle criterion for $\alpha = 0$ for the predictor slope step.

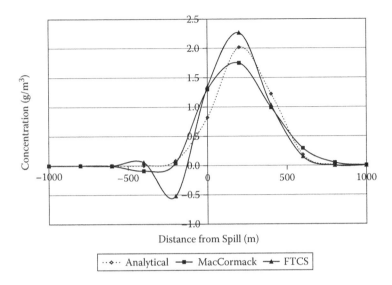

Distance from Spill (m)

···◇··· Analytical —■— MacCormack —▲— FTCS

FIGURE 6.16 Diminished wiggle from MacCormack method.

no-wiggle at $\Delta x = 100$. In addition, the numerical dispersion is relatively minor as one can see by comparing the MacCormack results to the analytical solution.

To further examine the relative susceptibility of the MacCormack method to wiggle versus the E-unmodified[*] FTCS method, we ran the spill problem at $\Delta x = 200$ m (which violates the no-wiggle criterion for both methods), $E_m = E_p$, and a time step of 0.8 days for both methods. The results at 2.4 days after the spill are shown in Figure 6.16. While the MacCormack results show some wiggle, it is far less than the result of the FTCS simulation.

6.4 IMPLICIT BACK TIME METHOD

6.4.1 METHODOLOGY

Unlike the forward time Euler method, the implicit back time method evaluates the right sides of the generalized system of Equations (4.20) or (4.22) at time $t + 1$. Therefore, instead of the Euler Equation (6.7) we have

$$\frac{(\mathbf{R}(\underline{c})\mathbf{V}\underline{c})^{t+1} - (\mathbf{R}(\underline{c})\mathbf{V}\underline{c})^{t}}{\Delta t} = (\mathbf{A}\underline{c} + \underline{f}(\underline{c}) + \underline{b})^{t+1} \qquad (6.58)$$

For linear systems, (6.58) becomes

$$\frac{(\mathbf{R}\mathbf{V}\underline{c})^{t+1} - (\mathbf{R}\mathbf{V}\underline{c})^{t}}{\Delta t} = (\mathbf{A}\underline{c} + \underline{b})^{t+1}$$

---

[*] We have already seen that the E_m-modified FTCS method can considerably increase the no-wiggle Δx and/or stable Δt by a suitable selection of β and γ. For example, from Table 6.1, β and $\gamma = 0.9$ would allow a no-wiggle Δx of 300 m. Thus, we are comparing the MacCormack method only to the E-unmodified FTCS method where the no-wiggle length cannot be so manipulated.

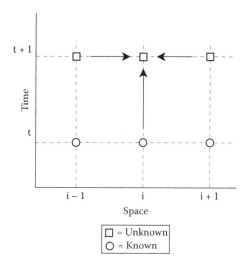

FIGURE 6.17   Time and space stencil for BTCS method.

and can be arranged as

$$((\mathbf{RV})^{t+1} - \Delta t \mathbf{A}^{t+1}) \underline{\mathbf{c}}^{t+1} = (\mathbf{RV}\underline{\mathbf{c}})^{t} + \Delta t \underline{\mathbf{b}}^{t+1} \tag{6.59}$$

In Equation (6.59), all the terms in the left side matrix of coefficients of $\underline{\mathbf{c}}^{t+1}$ (terms inside brackets) are presumed known at $t + 1$. On the right side, all terms also have known values because the concentrations are from the previous time step. Therefore, Equation (6.59) is a system of NEQN linear equations in NEQN unknowns (elements of $\underline{\mathbf{c}}^{t+1}$) and is solvable using Gaussian elimination techniques, as discussed in Chapter 5.

A time/space stencil for the BTCS method is shown in Figure 6.17. For nonlinear systems, Equation (6.58) is rewritten in an implicit form (set equal to zero) suitable to the GEM's quasi-Newton method as

$$\underline{\mathbf{g}}(\underline{\mathbf{c}}) = [\mathbf{R}(\underline{\mathbf{c}})\mathbf{V}\underline{\mathbf{c}}]^{t+1} - [\mathbf{R}(\underline{\mathbf{c}})\mathbf{V}]^{t} - \Delta t[\mathbf{A}\underline{\mathbf{c}} + \underline{\mathbf{f}}(\underline{\mathbf{c}}) + \underline{\mathbf{b}}]^{t+1} = \underline{\mathbf{0}} \tag{6.60}$$

where again it is understood that the elements of the $\underline{\mathbf{c}}$ vector at time $t + 1$ are the unknowns to be determined by the Newton algorithm for time step $t + 1$. Similar to but somewhat more complicated than the Euler case above, the i,jth element of the Jacobian matrix is

$$\frac{\partial g_i}{\partial c_j} = V_{ij} \frac{\partial \left[R_{ij}(\underline{\mathbf{c}}) c_i \right]^{t+1}}{\partial c_j} - \Delta t \left[A_{ij} + \frac{\partial f_i}{\partial c_j} \right]^{t+1} \tag{6.61}$$

where, as noted previously, $\partial f_i / \partial c_j$ is the partial derivative of the ith row of the nonlinear source/sink vector $\underline{\mathbf{f}}(\underline{\mathbf{c}})$ with respect to c_j. Also, identically to the Euler case above,

$$\frac{\partial\left[R_{ij}(\underline{\mathbf{c}})c_i\right]^{t+1}}{\partial c_j} = \left[R_{ij}(\underline{\mathbf{c}}) + c_i\frac{\partial R_{ij}(\underline{\mathbf{c}})}{\partial c_j}\right]^{t+1} \quad \text{if, } i = j \quad \text{(6.14a repeated)}$$

$$\frac{\partial\left[R_{ij}(\underline{\mathbf{c}})c_i\right]^{t+1}}{\partial c_j} = 0 \quad \text{if, } i \neq j \qquad \text{(6.15b repeated)}$$

Again, if analytical derivatives are used, the user supplies $\partial R_{ij}(\underline{\mathbf{c}})/\partial c_j$ and $\partial f_i/\partial c_j$, for all i and j.

6.4.2 STABILITY, NUMERICAL DISPERSION, EQUIVALENCE OF E-MODIFIED VERSIONS, AND WIGGLE

Stability and numerical dispersion characteristics are examined by simple analogies to the FT Euler methods. Considering stability, recall that the system of equations for the FT Euler is

$$\frac{(\mathbf{R}(\underline{\mathbf{c}})\mathbf{V}\underline{\mathbf{c}})^{t+1} - (\mathbf{R}(\underline{\mathbf{c}})\mathbf{V}\underline{\mathbf{c}})^{t}}{\Delta t} = (\mathbf{A}\underline{\mathbf{c}} + \underline{\mathbf{f}}(\underline{\mathbf{c}}) + \underline{\mathbf{b}})^{t} \qquad \text{(6.7 repeated)}$$

and that the Euler update for the ith compartment can be written for linear systems as

$$c_i^{t+1} = \frac{1}{R_{i,i}^{t+1}V_{i,i}^{t+1}}\left[\left(R_{i,i}^{t}V_{i,i}^{t} + \Delta t A_{i,i}^{t}\right)c_i^{t} + \Delta t\left[\sum_{j\neq i}A_{i,j}^{t}c_j^{t} + b_i^{t}\right]\right] \quad \text{(6.15 repeated)}$$

and that this update is stable if the coefficient of c_i^{t} is positive or

$$R_{i,i}^{t}V_{i,i}^{t} - \Delta t\left|A_{i,i}^{t}\right| > 0 \qquad \text{(6.17 repeated)}$$

leading to the following constraint on Δt:

$$\Delta t < \frac{R_{i,i}^{t}V_{i,i}^{t}}{\left|A_{i,i}^{t}\right|} \qquad \text{(6.18 repeated)}$$

The reason that the coefficient of c_i^{t} involves a term $(-\Delta t\left|A_{i,i}^{t}\right|)$ that can cause negativity is simply because the right side of Equation (6.7) was evaluated at time t in the FT Euler methods. In contrast, the BT methods evaluate the right side at time $t + 1$ [Equation (6.58)] and this simple difference (at least in this context) precludes the possibility that the coefficient of c_i^{t} will involve a term that can cause negativity. Thus, the BT methods are unconditionally stable, meaning that arbitrarily large time steps can be made without causing instability—a major advantage in many applications.

Considering numerical dispersion characteristics of the BT methods, recall that the temporal bias (non-centered time derivative approximations) for the FT methods resulted in a temporal numerical dispersion equal to

$$-\frac{\Delta t}{2R}\bar{U}^2$$

For the BT methods now under consideration, we are now conducting the bias in the opposite direction (backward instead of forward). It should not be surprising that the temporal dispersion has the same magnitude but is opposite in sign:

$$+\frac{\Delta t}{2R}\bar{U}^2$$

This result can be shown by a Taylor series analysis analogous to the FT examples. With regard to spatial numerical dispersion, it also should not be surprising that the BT and FT methods have identical characteristics, i.e., no spatial numerical dispersion for a centered spatial derivative approximation (BTCS and FTCS) and, for the BTBS and FTBS methods, a spatial numerical dispersion equal to

$$\bar{U}\frac{\Delta x}{2}$$

Thus, to summarize, the total numerical dispersion (E_n) results for the BT methods (and results of FT methods for comparison) are given in Table 6.2.

Recall for the FT methods that when E_n is controlled for, the FTBS and FTCS methods are equivalent. That equivalence arose because the modification to correct the FTBS method for spatial E_n transformed the FTBS method into the E-unmodified FTCS method. The further modification of both methods to correct for temporal E_n was identical because the temporal E_n term is identical for both methods. From inspection of Table 6.2, it is no surprise that this concept also applies to the E-modified

TABLE 6.2

Comparison of Total Numerical Dispersion (E_n) Results

FTCS	BTCS	FTBS	BTBS
$-\dfrac{\Delta t}{2R}\bar{U}^2$	$\dfrac{\Delta t}{2R}\bar{U}^2$	$\bar{U}\dfrac{\Delta x}{2}-\dfrac{\Delta t}{2R}\bar{U}^2$	$\bar{U}\dfrac{\Delta x}{2}+\dfrac{\Delta t}{2R}\bar{U}^2$

BTBS and BTCS methods. The only difference relative to the FT methods is that the temporal E_n has the opposite sign (caused by "back" time versus "forward" time), and is subtracted rather than added. After E modifications, they are equivalent.

Although both the BTBS and BTCS methods are unconditionally stable, after they are E-modified and become CS methods, wiggle remains an issue (as for all CS methods). Moreover, the vulnerability to wiggle considerably diminishes the unconditional stability advantage as shown below. From the no-wiggle criterion,

$$\Delta x < \frac{E_m}{0.5\overline{U}}$$

and, correcting for E_n using the BTCS E_n,

$$E_m = E_p - \frac{\Delta t \overline{U}^2}{2R}$$

Substituting the corrected E_m into the no-wiggle criterion, we have the no-wiggle Δx as a function of the system parameters E_p, $\overline{U}$, R, and the desired time step as

$$\Delta x < \frac{E_p - \dfrac{\Delta t \overline{U}^2}{2R}}{0.5\overline{U}} = \frac{2E_p}{\overline{U}} - \overline{U}\frac{\Delta t}{R} \tag{6.62a}$$

or, if we are interested in the maximum time step for a given Δx, we can solve (6.62a) for Δt as

$$\Delta t < R\left(\frac{2E_p - \overline{U}\Delta x}{\overline{U}^2}\right) \tag{6.62b}$$

Equation (6.62) is somewhat analogous to Equation (6.46) for the E-modified FT methods. It shows that wiggle is affected by *both* Δx and Δt. If we want a larger wiggle-free Δx, we must decrease Δt and vice versa.

This allowable time step (6.62b) for the E-modified BT methods can be more limited by wiggle than the allowable time step for the FT methods is limited by stability, thus diminishing the unconditional stability advantage of the BT methods. For example, returning to our aquifer spill example ($\overline{U} = 160$, $E_p = 1.0 \times 10^4$, $R = 1.68$, $k = -0.1$, and $\Delta x = 100$), Equation (6.62b) gives a maximum no-wiggle time step of 0.263 days. In contrast, the maximum time step for stability for the E-modified FT methods (they allow greater time steps than the E-unmodified FT methods) from Equation (6.46) with $\gamma = 1.0$ (0 safety factor for stability) yields $\Delta t = 0.567$ days or more than twice the result from the no-wiggle BT method with E modification.

Conversely, if we wanted to maintain the 0.567-day time step, the maximum no-wiggle Δx from Equation (6.62a) is only 71 m as opposed to 125 m for the BT method.

EXAMPLE 6.7

BT versus FT Methods

We made several GEM runs of the aquifer spill example to illustrate the advantages and disadvantages of the BT methods versus the FT* methods.

The chief advantage of the E-unmodified BT method compared to the FT method is its unconditional stability. Figure 6.18 shows results of the E-modified FT method applied to our aquifer spill problem (the FTCS method was used with $E_m = 1.43 \times 10^4$) at a time step of 0.6 days, which very slightly exceeds its maximum for stability of 0.567 days. We also ran the E-unmodified BTBS and BTCS methods at a time step of 3.6 days (six times greater). While the FT method is wildly unstable, both BT methods are stable and wiggle-free, albeit with numerical dispersion and some considerable discretization error due to the large time step as can be seen by comparison to the analytical solution. $\Delta x = 100$ was used for all runs and $E_p = 1 \times 10^4$, R = 1.68, $\overline{U} = 160$, and $k = -0.1$ were again assumed. The simulations are shown at time 3.6 days (representing one time step for the BT methods and six time steps for the FTCS methods). The BTBS and BTCS results are

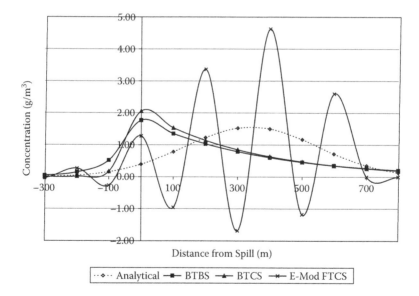

FIGURE 6.18   Unconditional stability of BT methods.

---

* The FT and BT methods are very similar in concept but exhibit remarkably different numerical characteristics as we have seen. That is why we focus on their comparison. More general comparisons across all methods are included in a later section.

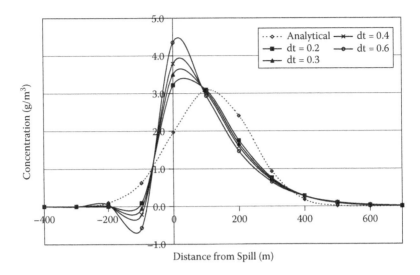

Distance from Spill (m)

FIGURE 6.19 Increasing susceptibility to wiggle of BT methods.

similar but not identical. Although the BT results are far from accurate because of the discretization error and E_n, the point is that they are stable and wiggle-free.

We next made several runs of the BT methods at different time steps after modifying E_m to illustrate the increasing effects of wiggle. These results (all at time 1.2 days after the spill) are shown in Figure 6.19. Again $\Delta x = 100$ and $E_p = 1 \times 10^4$, $R = 1.68$, and $k = -0.1$ were used. As noted above, the maximum Δt for no-wiggle is 0.263 days. We made a BTCS run at $\Delta t = 0.2$ days (six time steps) and $E_m = 8.47 \times 10^3$. (An equivalent run would be BTBS with $E_m = 473.0$.) We also made BTCS runs at $\Delta t = 0.3$ days (four time steps), 0.4 days (three time steps), and 0.6 days (two time steps). The respective corrected E_m values are 7.71×10^3, 6.94×10^3, and 5.41×10^3. BTBS runs at the same time steps and corrected E_m values of -273.0, -1057.0, and -2585.0, respectively, yielded identical results. All three time steps exceeded the no-wiggle criterion.

Although none of the simulations involved numerical dispersion, the implications of exceeding the no-wiggle criterion become increasingly severe as the time step increases beyond the no-wiggle maximum. The errors illustrated in the figure represent wiggles resulting from the E_m correction. They are not instabilities.

The conclusion is that, while the unconditionally stable BT methods allow larger time steps than the conditionally stable FT methods, eliminating numerical dispersion quickly results in wiggle errors. Therefore, while the implicit BT method can certainly provide a relative advantage in terms of allowable time step sizes,* to realize this advantage you must accept numerical dispersion using the E-unmodified BT methods.

---

* As noted by Chapra (1997), even though these BT methods are subject to significant error at large time steps, they are often used advantageously to solve steady-state problems by running a dynamic problem out to steady state using large stable time steps. The GEM has *direct* methods for solving steady-state problems so this advantage, while worth noting, is somewhat moot for our purposes.

6.5 IMPLICIT CENTERED TIME (CRANK–NICOLSON) METHOD

6.5.1 METHODOLOGY

This final GEM option is an implicit centered time method. When a spatial central difference is used ($\alpha = 0.5$), the option is also known as the Crank–Nicolson (1947) method. Unlike the forward time Euler and center time MacCormack explicit methods, it is unconditionally stable (like the back time method) so that larger time steps can be used. Unlike the forward time Euler and back time methods, it uses a temporal central difference so that temporal numerical diffusion will be eliminated. The centered time method incorporates advantageous aspects from both the back time and MacCormack methods into a single, one-step[*] method. At time step t, the centered time method is

$$\frac{(\mathbf{R}(\underline{c})\mathbf{V}\underline{c})^{t+1} - (\mathbf{R}(\underline{c})\mathbf{V}\underline{c})^{t}}{\Delta t} = \frac{[(\mathbf{A}\underline{c} + \underline{f}(\underline{c}) + \mathbf{b})^{t} + (\mathbf{A}\underline{c} + \underline{f}(\underline{c}) + \mathbf{b})^{t+1}]}{2} \tag{6.63}$$

which the observant reader will recognize as identical to Equation (6.55) after rearrangement presented in the development of the MacCormack method. In other words, like the MacCormack method, the centered time method evaluates the slope ($\mathbf{A}\underline{c} + \underline{f}(\underline{c}) + \mathbf{b}$) at both time t and time $t + 1$ and uses the average. However, unlike the MacCormack method (and like the back time method), the slope at $t + 1$ is not *estimated* by a predictor step, but rather is determined implicitly in a single step by solving the system of simultaneous equations. The fact that the slope at $t + 1$ is not estimated, but rather fully determined, makes this a fully time-centered method.

A time/space stencil of the CTCS method is shown in Figure 6.20. It should be noted that the time/space stencil, and the "centered-time" name itself, may seem to suggest that the updated concentration predictions at each time step are relevant to the time step *midpoint*, i.e., $t + \Delta t/2$. This is not the case. The concentration predictions are applicable to the end of the full time step, i.e., $t + \Delta t$, just as in every other method we have discussed. What the "centered-time" strategy has done is to make these full time step predictions using *spatial* estimates of the right-hand side (recall our Section 6.1 Introduction discussion) that correspond to the temporal midpoint. Using these temporal midpoint estimates, we are making full time step predictions.

For linear systems, the centered time method is written in the form [compare to Equation (6.59) for the back time method]

$$\left((\mathbf{RV})^{t+1} - \frac{\Delta t}{2}\mathbf{A}^{t+1} \right)\underline{c}^{t+1} = (\mathbf{RV}\underline{c})^{t} + \frac{\Delta t}{2}\left(\underline{b}^{t+1} + \mathbf{A}\underline{c}^{t} + \underline{b}^{t} \right) \tag{6.64}$$

where the unknowns $\underline{c}^{t+1}$ are on the left side and all knowns are on the right side, amenable to the linear algebra solution. For nonlinear problems, it is written in implicit form, amenable to the quasi-Newton algorithm, as

---

[*] Recall that the MacCormack is a two-step method.

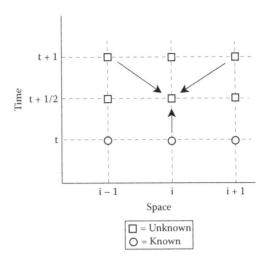

FIGURE 6.20   Time and space stencil for CTCS method.

$$\underline{g}(\underline{c}) = [\mathbf{R}(\underline{c})\mathbf{V}\underline{c}]^{t+1} - [\mathbf{R}(\underline{c})\mathbf{V}]^{t} - \frac{\Delta t}{2}\left\{[\mathbf{A}\underline{c} + \underline{f}(\underline{c}) + \underline{b}]^{t+1} + [\mathbf{A}\underline{c} + \underline{f}(\underline{c}) + \underline{b}]^{t}\right\} = \underline{0} \quad (6.65)$$

where again it is understood that the elements of the $\underline{c}$ vector at time $t + 1$ are the unknowns to be determined by the Newton algorithm for time step $t + 1$.

Nearly identical to the implicit BT case above, the i,jth element of the Jacobian matrix is

$$\frac{\partial g_i}{\partial c_j} = V_{ij}\frac{\partial[R_{ij}(\underline{c})c_i]^{t+1}}{\partial c_j} - \frac{\Delta t}{2}\left[A_{ij} + \frac{\partial f_i}{\partial c_j}\right]^{t+1} \quad (6.66)$$

where, as noted previously, $\partial f_i/\partial c_j$ is the partial derivative of the ith row of the non-linear source/sink vector $\underline{f}(\underline{c})$ with respect to c_j. Also identical to the FT and BT cases above,

$$\frac{\partial[R_{ij}(\underline{c})c_i]^{t+1}}{\partial c_j} = [R_{ij}(\underline{c}) + c_i\frac{\partial R_{ij}(\underline{c})}{\partial c_j}]^{t+1} \text{ if, } i = j \quad \text{(6.14a repeated)}$$

$$\frac{\partial[R_{ij}(\underline{c})c_i]^{t+1}}{\partial c_j} = 0 \text{ if, } i \neq j \quad \text{(6.14b repeated)}$$

Again, if analytical derivatives are used, the user supplies $\partial R_{ij}(\underline{c})/\partial c_j$ and $\partial f_i/\partial c_j$, for all i and j.

6.5.2   STABILITY, NUMERICAL DISPERSION, EQUIVALENCE
OF E-MODIFIED METHODS, AND WIGGLE

We were able to show the unconditional stability of the BT methods by a relatively simple analogy to the FT stability analysis. That approach becomes more difficult for the CT methods and we will simply state[*] that the methods (both CTCS and CTBS) are unconditionally stable (Hoffman, 1992; Chapra, 1997).

Regarding numerical dispersion, the CTCS (Crank–Nicolson) method is a true, time-centered, space-centered method. We already know that means no temporal or spatial numerical dispersion. The CTBS method involves spatial numerical dispersion and it can be shown to be $E_n = +U\Delta x/2$, i.e., like any other BS method.

Once again, and for the reasons previously discussed, if the CTBS method is E modified to correct for its numerical dispersion, it becomes equivalent to the CTCS method. We might not want to correct for the CTBS method for E_n because, as a BS method, it is not subject to wiggle whereas the CTCS method (and the E-modified CTBS method) is. In a situation where very large compartment sizes are desired and a user is willing to live with the numerical dispersion, the E-unmodified CTBS method may be the better choice.

EXAMPLE 6.8

CT Methods

We made several GEM runs to illustrate the CT methods. We continue to use the aquifer spill problem ($\Delta x = 100$, $E_p = 1 \times 10^4$, $\overline{U} = 160$, R = 1.68, and $k = -0.1$) for continuity and the availability of an analytical solution for comparison.

We first ran the CTCS and CTBS options at a 0.8-day time step. This was the maximum time step that satisfied stability for the FTCS method (and we previously used 0.6 days to be conservative). It also significantly violates the no-wiggle time step for the E-modified BT methods. Three time steps were used and the profiles are shown in Figure 6.21 at 2.4 days after the spill. No correction for numerical dispersion was made for the CTBS option. (None is needed for CTCS.) Although the CTCS solution slightly overpredicts the peak concentration, due to temporal discretization error from the relatively large time step, it generally has excellent agreement with the analytical solution, demonstrating its accuracy and the absence of numerical dispersion. The CTBS solution shows numerical dispersion.

We next ran the CTCS option at three progressively larger time steps (1, 2, and 4 days) to demonstrate unconditional stability relative to the FT methods. Three time steps were made for each run so the three runs indicate spill progression downstream. Figure 6.22 shows the results along with the corresponding analytical solutions. (E_m-modified CTBS options give the same results.) What we see is that at time steps exceeding 1 day, although the peak concentrations are simulated reasonably accurately even at the 4-day time step, significant inaccuracy appears upstream of the peak.

---

[*] Hoffman (1992) presents an excellent overview of three different mathematical methods used to analyze finite difference equations for stability: (1) the discrete perturbation method, (2) the von Neumann method, and (3) the matrix method. The Courant condition (see Box 6.1) previously discussed for the FT methods would result from any of these.

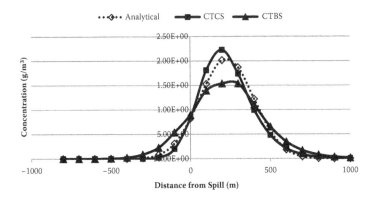

FIGURE 6.21 CT results.

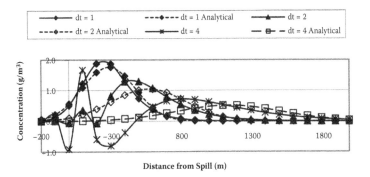

FIGURE 6.22 Example of enhanced stability of CT relative to FT methods.

These errors appear as wiggle but are in fact not wiggle (at least as we define it as exceeding the no-wiggle Peclet criterion), because the $\Delta x = 100$ m spatial discretization more than satisfies the no-wiggle criterion. Rather, these oscillations occur with the CT method when it is applied to problems involving significant boundary or initial condition transients or discontinuities.

We have not previously discussed these transient oscillations because they are apparently unique to the CT method, but they demand some attention here. The transient/discontinuity in our problem is associated with the initial condition—the spill. By definition, a spill introduces a significant transient/discontinuous condition into a simulation problem. Criteria for avoiding these transient oscillations have been developed by Hossain and Miah (1999). These criteria involve the Peclet (Pe) and Courant (Cr) numbers, which were introduced previously. The Peclet number is defined as

$$Pe = \frac{\bar{U}_{ij}L_{ij}}{E_{ij}} \qquad (6.67)$$

and the Courant number as

$$Cr = \frac{\bar{U}_{ij}\Delta t}{R_{ij}E_{ij}} \qquad (6.68)$$

For continuous sources, Hossain and Miah give the following criteria:

$$Pe \leq 50, \; PeCr \leq 5, \; \text{and} \; Cr \leq 1 \qquad (6.69)$$

For "pulse" sources, the following more restrictive criteria apply:

$$Pe \leq 20, \; PeCr \leq 5, \; \text{and} \; Cr \leq 1 \qquad (6.70)$$

Using the pulse criteria to reflect our spill problem and, considering the 2-day time step at which we first noticed oscillations in Figure 6.22, we see that

$$Pe = \frac{\overline{U}_{ij}L_{ij}}{E_{ij}} = \frac{(160)(100)}{10,000} = 1.6, \; \text{which is OK}$$

$$Cr = \frac{\overline{U}_{ij}\Delta t}{R_{ij}E_{ij}} = \frac{(16)(2)}{(1.68)(10,000)} = 1.91, \; \text{which violates the criterion}$$

Notice, however, that a 1-day time step does (barely) satisfy the Cr criterion at 0.95 and, indeed, we did not see oscillations with this time step in Figure 6.22.

These oscillations, while obviously problematic for simulating conditions near the transients in time and space, decay over time and become less of an issue. To illustrate this decay, we ran the CTCS method using the 2-day time step to simulate profiles at times 5, 7, 11, and 15 days after the spill. Figure 6.23 shows the diminution of the transient oscillations over time. Again, these oscillations do not represent wiggle and are different phenomena.

Finally, returning to excessive spatial disaggregation-caused wiggle, we made runs using $\Delta x = 200$ m to illustrate the CTCS method's susceptibility to wiggle and the no-wiggle advantage of E-unmodified CTBS. Results are shown in Figure 6.24. A time step of 0.8 days was used for both runs and the results are shown after three time steps at 2.4 days after the spill. Although the CTBS results again show numerical dispersion, they are wiggle-free whereas the CTCS results exhibit wiggle.

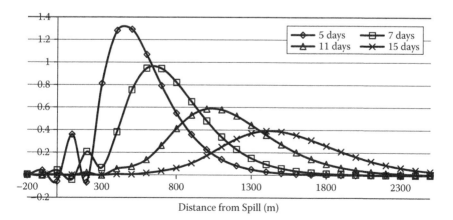

FIGURE 6.23 Decaying CTCS transient oscillations.

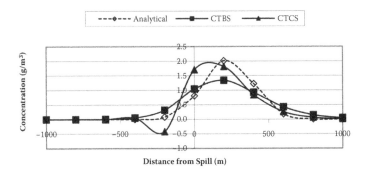

FIGURE 6.24 Wiggle versus E_n.

6.6 CROSS-METHOD COMPARISONS AND SUMMARY OF ERROR CHARACTERISTICS FOR LINEAR PROBLEMS

6.6.1 Cross-Method Comparisons

In the preceding Sections 6.2 through 6.5, although a few cross-method comparisons were made to demonstrate certain features, the comparisons were not inclusive of all methods or criteria that may indicate relative performance. A more complete cross-method set of comparisons follows. The comparisons continue to use the one-dimensional aquifer spill problem ($E_p = 1 \times 10^4$, $\overline{U} = 160$, $R = 1.68$, and $k = -0.1$).

6.6.1.1 Maximum Time Step

First, we consider which method allows the largest reasonably accurate time step for a fixed Δx, and assume $\Delta x = 100$ m as earlier. We require no numerical dispersion and wiggle may not be excessive.

For the FT methods, given our zero numerical dispersion requirement, we consider only the E-modified versions. To conservatively maximize the stable time step, we used $\gamma = 0.8$ for the stable time step safety factor. The corresponding time step is 0.476 days. For the FTCS method, $E_m = 1.36 \times 10^4$. Recall our maximum Δx for no wiggle with $E_m = E_p$ is 125 m. Because $E_m = E_p + E_n$, our E-modified method increases the no-wiggle Δx so 100 m is OK.

For the MacCormack method, we have seen that 100 m violates the no-wiggle criterion, because of the $\alpha = 0$ predictor step, but we included this method because it exhibits some robustness to wiggle violations. Regarding a stable time step, we first tried the same Δt as for the FT (0.476) and noted quite good results with very minor wiggle. We then increased Δt by 2 to 0.714 and still got relatively good results although wiggle and/or instability began to appear. Above that Δt, wiggle and/or instability were prohibitive. As usual for this method, $E_m = E_p$.

Although the BT methods are unconditionally stable, the zero numerical dispersion requirement forced us to use the E-modified versions where the time step, although stable, can cause wiggle. Indeed, because we subtract E_n from E_p to get the modified E_m, E_m becomes smaller than E_p as we increase Δt above 0. Recall that the maximum Δx for $E_m = E_p = 1.0 \times 10^4$ for no wiggle was 125 m (we have been using 100 as a safety factor). Therefore, even at a Δt of 0, our no-wiggle maximum Δx is only 125. Nonetheless, we tried the same Δt as used to meet stability for the FT methods (0.476 days). As expected, minor wiggle resulted. The results were otherwise excellent.

For the CT methods, we used the CTCS. This method is unconditionally stable and the no-wiggle criterion is only a function of Δx, not Δt, so wiggle is not an issue in selecting a time step. However, as shown previously, decaying oscillations can occur in the neighborhood of transient or discontinuous conditions as we saw at time steps exceeding approximately 1 day that violate the Hossain and Miah oscillation criteria. Nonetheless, we found by trial and error that a time step of 1.5 days does not exhibit excessive oscillation even after a single time step.

In summary, the CTCS method clearly allows the relatively larger time step while incurring no numerical dispersion or wiggle and only minor oscillation errors. The MacCormack method is also relatively good in this comparison.

6.6.1.2 Maximum Compartment Volume

The second cross-method comparison considers which method allows the largest wiggle-free Δx for a fixed Δt with no numerical dispersion in a stable solution.

We can immediately eliminate the MacCormack and CT methods from this comparison because their no-wiggle maximum compartment lengths are determined exclusively by E_p (125 m for the CT and somewhat less for MacCormack), regardless of the time step. The E-modified BT method is even worse because its maximum no-wiggle length is a function of both E_p and Δt and the time step works against the maximum length. In other words, for any $\Delta t > 0$, the maximum length is <125 so it is eliminated as well. The E-unmodified BT methods are also limited to 125 m and show numerical dispersion.

We have already seen that the E-modified FT methods allow a significant increase in Δx relative to the other methods; however, before we simply declare an E-modified FT method a winner, we must make sure that its Δx increase (above 125 m) does not come at the expense of a smaller Δt than the other methods. (Remember, we are comparing maximum length for a fixed time step.) The benchmark Δt would then seem to be the CTCS method value of approximately 1.5 days from the preceding comparison.

At that time step and $\Delta x = 100$ m, the CTCS result was marginally acceptable. Thus, we are asking whether at $\Delta t = 1.5$ days, the E-modified FT method can produce an acceptable result at any $\Delta x > 100$ to 125 m. Using Equation (6.46) we searched for a combination of γ and β (assuming they are equal to make this problem tractable) that would be reasonably conservative and yield a time step of approximately 1.5 days. By trial and error, we found that $\gamma = \beta = 0.84$ yields $\Delta t = 1.48$ days. The corresponding $\Delta x = 224$ m; therefore, this method is the winner.

6.6.1.3 Pure Advection ($E_p = 0$)

We are interested here in which methods are best or even feasible for pure advection problems. First, we can dismiss CS methods that use $E_m = E_p = 0$, because they are unconditionally wiggly, i.e.,

$$\Delta x < \frac{E_m}{0.5\overline{U}} < 0.$$

In our discussion of FT methods, we noted that the FTBS method with $E_m = E_p = 0$ seems ideally suited to the pure advection problem because we can both maximize the time step and eliminate numerical dispersion. Given this candidate for the best pure advection method, the question then becomes whether another method can use an even larger time step and also avoid numerical dispersion, wiggle, and the downstream boundary condition problem.

Let's consider the MacCormack method. Although its $\alpha = 0$ predictor step suggests that it is even more vulnerable than a CS method to (1) wiggle because $E_m = E_p = 0$ and (2) the downstream boundary condition problem, we were impressed by the robustness of the MacCormack method so it is worth consideration. We ran the method at the same time step (0.98529 days) as the maximum time step for the E-unmodified FTBS method. (The downstream boundary condition problem is not an issue for our spill problem because we know it is zero at least until the plug arrives at the downstream compartment.) We did not see evidence of wiggle, but numerical dispersion made the MacCormack method unsuitable.

The BT methods are unconditionally stable, so this suggests a possibility of a larger time step. Because accurate simulation of the plug flow spill problem requires no numerical dispersion, we are limited to consideration of the E-modified BT methods. That presents an immediate wiggle problem, because the numerical dispersion for both the E-modified BTBS and BTCS methods is a positive value (unlike the FT methods); thus, we must *subtract* E_n from E_p (0), resulting in negative E_m and unconditional wiggle. Similarly, the CT methods, although also unconditionally stable, involve either $E_m = 0$ (CTCS) or $E_m = E_p - E_n = 0 - E_n$ (CTBS with E modification to eliminate numerical dispersion) and will be unconditionally wiggly.

In conclusion, the E-unmodified FTBS method using the maximum stable time step is the only viable method for pure advection problems.

6.6.1.4 Pure Dispersion ($Q = 0$)

This cross-method comparison looks at a pure dispersion problem, i.e., flow and velocity are 0s. The other parameters of our spill example remain the same. Like the pure advection problem, our interest here is the extent to which the four methods are applicable, and whether any method seems preferable.

We first note that numerical dispersion for all methods involving numerical dispersion is caused by advective flow. For zero flow, there is zero numerical dispersion. This is true for both spatial and temporal numerical dispersion. Therefore, for a pure dispersion problem, we have no need to correct methods for numerical dispersion. The FT methods that are not unconditionally stable present identical stability criteria. Indeed,

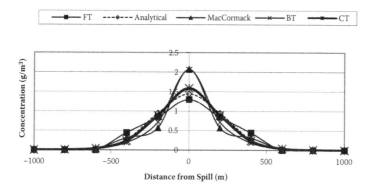

FIGURE 6.25 Comparison of methods for pure dispersion problem.

because there is no advective flow, the spatial differencing parameter for advective flow α is moot and the FTBS and FTCS methods are identical. This is true for BTBS and BTCS and CTCS and CTBS as well. We also note that all methods are wiggle-free because the no-wiggle criterion divides the dispersion coefficient by the velocity (0) resulting in an infinite no-wiggle compartment length. Thus, all methods involve zero numerical dispersion and are wiggle-free. The only difference is that the FT and MacCormack methods have stability limits and the BT and CT methods do not.

To compare the four methods, we used a compartment length of 200 m and a time step of

$$\Delta t < \frac{R\Delta x^2}{2E_p + k\Delta x^2} = \frac{1.68(200)^2}{2(10,000) + 0.1(200)^2} < 2.8,$$

say, 2.0 days to be conservative for the FT and MacCormack methods. To account for the twice-larger compartment volumes, we halved the initial condition. Also, because there is no advective flow, we moved the loading farther downstream from the previous compartment 10 to compartment 50 so that we could safely assume zero boundary conditions for both upstream and downstream boundary compartments (1 and 100). $E_m = E_p = 1 \times 10^4$ for all methods. Two time steps were made and the results of the FT, MacCormack, BT, and CT methods and the analytical solution are shown in Figure 6.25 at 4 days after the spill.

Although all methods performed reasonably well at this relatively large compartment size and time step (compared to previous runs), the FT and CT methods are the most accurate—at least for this example. The BT and MacCormack methods somewhat overestimated the peak concentration.

6.6.2 Summary of Error Characteristics

A summary of the error characteristics along with the advantages and disadvantages of the four methods and sub-methods follows for linear systems.

6.6.2.1 Forward Time Explicit (Euler) Methods

$E_p > 0$, $\overline{U} > 0$ (advective/dispersive problems)

$E_m = E_p$ (E-unmodified)

FTBS (Δx unconditionally wiggle-free)

$$\Delta t < \frac{R\Delta x^2}{\left|-\overline{U}\Delta x - 2E_p \pm k\Delta x^2\varphi^{T-20}\right|} \quad \text{for stability}$$

$$E_n = \frac{\overline{U}\Delta x}{2} - \frac{\Delta t\overline{U}^2}{2R}$$

FTCS

$$\Delta x < \frac{2E_p}{\overline{U}} \quad \text{for no wiggle}$$

$$\Delta t < \frac{R\Delta x^2}{\left|-2E_p \pm k\Delta x^2\varphi^{T-20}\right|} \quad \text{for stability}$$

$$E_n = -\frac{\Delta t\overline{U}^2}{2R}$$

$E_m = E_p \pm E_n$ (E-modified; FTBS and FTCS are equivalent)

For zero numerical dispersion,

$$\text{FTBS} \quad E_m = E_p - \frac{\overline{U}\Delta x}{2} + \frac{\Delta t\overline{U}^2}{2R}$$

For zero numerical dispersion,

$$\text{FTCS} \quad E_m = E_p + \frac{\Delta t\overline{U}^2}{2R}$$

Stable Δt with no wiggle and 0 E_n from solving quadratic equation

$$\Delta t^2\left[\pm\frac{\overline{U}^2|k|\beta^2\phi^{T-20}}{R}\right] + \Delta t\left[\overline{U}^2 \pm 2E_p|k|\beta^2\phi^{T-20} - \overline{U}^2\beta^2\gamma\right] - 2E_pR\beta^2\gamma = 0$$

where $|k|$ means using absolute value of k and $+$ sign if k is a sink (<0) or $-$ sign if a source (>0); $0 < \gamma$ and $\beta \leq 1$ and $\beta^2\gamma < 1$ are user-supplied safety factors for stability and no wiggle, respectively.

Stable and wiggle-free Δt and/or Δx can be increased up to limits on γ and β; alternatively, stable Δt with 0 E_n from solving quadratic equation

$$\frac{\overline{U}^2}{R}\Delta t^2 + (2E_p \pm |k|\Delta x^2 \varphi^{T-20})\Delta t - \gamma R \Delta x^2 = 0$$

for user-specified Δx and γ where again $|k|$ means absolute value of k, + symbol if k is a sink (<0) or – symbol if a source (>0).

$E_p = 0$ (pure advection problems)

E-unmodified FTBS ($E_m = E_p = 0$) with maximum stable Δt has no numerical dispersion and is wiggle-free.

$\overline{U} = 0$ (pure dispersion problems)

FTCS and FTBS methods are equivalent because α parameter is irrelevant.

No numerical dispersion for FTBS and FTCS because E_n is caused by $\overline{U}$.

$$\Delta t < \frac{R\Delta x^2}{\left|-2E_p \pm k\Delta x^2 \varphi^{T-20}\right|}$$

for stability for FTBS/FTCS.

FTBS and FTCS are unconditionally wiggle-free.

6.6.2.2 MacCormack Explicit Method

$E_p > 0$, $\overline{U} > 0$ (advective/dispersive problems)

Stability criterion has not been quantified and MacCormack method maximally stable Δt is somewhat greater than for E-unmodified FT methods.

No-wiggle Δx is also not quantified. For the predictor ($\alpha = 0$) step, the maximum no-wiggle Δx is half that of a CS ($\alpha = 0.5$) method. However, the corrector ($\alpha = 1$) step is unconditionally wiggle-free, so wiggle introduced during the predictor step will be mitigated, but not eliminated.

The MacCormack is not a true center time, center space method so some numerical dispersion occurs. It has not been quantified, but is less than dispersion for E-unmodified FT methods.

$E_p = 0$ (pure advection problems)
Not appropriate because of susceptibility to wiggle and numerical dispersion.

$\overline{U} = 0$ (pure dispersion problems)
No numerical dispersion because E_n is caused by $\overline{U}$.
Unconditionally wiggle-free.

6.6.2.3 Back Time Implicit Methods

$E_p > 0$, $\overline{U} > 0$ (advective/dispersive problems)

$E_m = E_p$ (E-unmodified)

BTBS

Δt unconditionally stable.

Δx unconditionally wiggle-free.

$$E_n = \frac{\overline{U}\Delta x}{2} + \frac{\Delta t \overline{U}^2}{2R}$$

BTCS

Δt unconditionally stable.

$$\Delta x < \frac{2E_p}{\overline{U}} \text{ for no wiggle.}$$

$$E_n = \frac{\Delta t \overline{U}^2}{2R}$$

$E_m = E_p - E_n$ (E-modified)

BTBS and BTCS are equivalent

BTBS $E_m = E_p - \dfrac{\overline{U}\Delta x}{2} - \dfrac{\Delta t \overline{U}^2}{2R}$

BTCS $E_m = E_p - \dfrac{\Delta t \overline{U}^2}{2R}$

No numerical dispersion
Unconditionally stable, but:

$$\Delta x < \frac{2E_p}{\overline{U}} - \overline{U}\frac{\Delta t}{R} \text{ for no wiggle given } \Delta t$$

$$\Delta t < R\left(\frac{2E_p - \overline{U}\Delta x}{\overline{U}^2}\right) \text{ for no wiggle given } \Delta x$$

$E_p = 0$ (pure advection problems)

BT methods are not suitable.
For no numerical dispersion, E_m will be negative for both BTBS and BTCS methods, leading to unconditionally wiggly Δx

$\overline{U} = 0$ (pure dispersion problems)

BTCS and BTBS methods are equivalent because α parameter is irrelevant.
No numerical dispersion for BTBS and BTCS because E_n is caused by $\overline{U}$.
BTBS and BTCS are unconditionally wiggle-free and unconditionally stable.

6.6.2.4 Center Time Implicit Methods

$E_p > 0$, $\overline{U} > 0$ (advective/dispersive problems)

$E_m = E_p$ (E-unmodified)

CTBS

Δt unconditionally stable.

Δx unconditionally wiggle-free.

$$E_n = \frac{\overline{U}\Delta x}{2}$$

Subject to transient or discontinuity-related oscillations at relatively large time steps.[*]

CTCS

Δt is unconditionally stable.

$\Delta x < \dfrac{2E_p}{\overline{U}}$ for no wiggle.

No numerical dispersion occurs.

Subject to transient or discontinuity-related oscillations at relatively large time steps.[*]

$E_m = E_p - E_n$ (E-modified)

Relevant only for CTBS.

CTBS becomes equivalent to CTCS as described above.

CTBS $E_m = E_p - \dfrac{\overline{U}\Delta x}{2}$

Subject to transient or discontinuity-related oscillations at relatively large time steps.[*]

$E_p = 0$ (pure advection problems)

CTCS is not suitable because $E_m = E_p = 0$, Δx is unconditionally wiggly.
CTBS is not suitable. If E_n is corrected, E_m is negative and CTBS equivalent to CTCS hence Δx unconditionally wiggly. If E_n is not corrected, numerical dispersion results.

$\overline{U} = 0$ (pure dispersion problems)

CTCS and CTBS methods are equivalent because α parameter is irrelevant.
No numerical dispersion for CTCS and CTBS because E_n is caused by $\overline{U}$.

---

[*] Oscillations may be avoided by selecting spatial and temporal discretizations that satisfy the Peclet and Courant-based criteria: Pe ≤ 50, PeCr ≤ 5, and Cr ≤ 1 for continuous sources and Pe ≤ 20, PeCr ≤ 5, and Cr ≤ 1 for pulse sources.

CTCS and CTBS are unconditionally wiggle-free and unconditionally stable.

Subject to transient or discontinuity-related oscillations at relatively large time steps.[*]

REFERENCES

Carnahan, B., Luther, H.A., and Wilkes, J.O. 1969. *Applied Numerical Methods*. New York: Wiley.

Chapra, S.C. 1997. *Surface Water-Quality Modeling*, preliminary ed. New York: McGraw-Hill.

Hoffman, J.D. 1992. *Numerical Methods for Engineers and Scientists*. New York: McGraw-Hill.

Hossain, M.A. and A.S. Miah. 1999. Crank–Nicolson–Galerkin Model for Transport in Groundwater: Refined Criteria for Accuracy, Applied Mathematics and Computation. *Applied Mathematics and Computation*, 105(2–3), 173–181.

MacCormack, R.W. 1969. *The Effect of Viscosity in Hypervelocity Impact Cratering*. American Institute of Aeronautics and Astronautics, Paper 69-354. AIAA Hypervelocity Impact Conference, Cincinnati, Ohio, April 30–May 2.

Remson, I., Hornberger, G.M., and Molz, F.J. 1971. *Numerical Methods in Subsurface Hydrology*. New York: Wiley-Interscience.

Schnoor, J.L. 1996. *Environmental Modeling: Fate and Transport of Pollutants in Water, Air, and Soil*. New York: Wiley.

Vestegaard, K. 1989. *Numerical Modelling of Streams: Hydrodynamic Models and Models for Transport and Spreading of Pollutants*. (Series Paper 1). Hydraulics & Coastal Engineering Laboratory, Department of Civil Engineering. Denmark: Aalborg University, June.

---

[*] Oscillations may be avoided by selecting spatial and temporal discretizations that satisfy the Peclet and Courant-based criteria: $Pe \leq 50$, $PeCr \leq 5$, and $Cr \leq 1$ for continuous sources and $Pe \leq 20$, $PeCr \leq 5$, and $Cr \leq 1$ for pulse sources

Appendix: Introduction to Matrices and Matrix Operations

The following material is excerpted and slightly modified from Wikipedia.com [http://en.wikipedia.org/wiki/Matrix_(mathematics)]. Wikipedia and many other references have much more information on matrix operations for the interested reader.

A *matrix* is a rectangular arrangement of numbers. For example

$$A = \begin{bmatrix} 9 & 8 & 6 \\ 1 & 2 & 7 \\ 4 & 9 & 2 \\ 6 & 0 & 5 \end{bmatrix}$$

The horizontal and vertical lines in a matrix are called *rows* and *columns*, respectively. The numbers in the matrix are called its *entries* or its *elements*. To specify matrix size, a matrix with m rows and n columns is called an $m \times n$ matrix, while m and n are called its dimensions. The above is a 4×3 matrix.

A matrix with one row ($1 \times n$ matrix) is called a *row vector* and a matrix with one column (an $m \times 1$ matrix) is called a *column vector*.

Matrices are usually denoted using upper-case letters, while the corresponding lower-case letters, with two subscript indices, represent the elements. In addition to using upper-case letters to symbolize matrices, many authors use a special typographical style, commonly boldface, to further distinguish matrices from other variables.

The element that lies in the ith row and the jth column of a matrix is typically referred to as the i,j, (i,j), or (i,j)th element of the matrix. For example, the (2,3) element of the above matrix A is 7. The (i,j)th element of a matrix A is most commonly written as $a_{i,j}$. Alternative notations for that entry are $A[i,j]$ or $A_{i,j}$.

A common shorthand is

$$A = A_{i,j} \text{ for } 1, \ldots, m; j = 1, \ldots, n$$

to define an $m \times n$ matrix A.

There are a number of operations that can be applied to modify matrices. They are called *matrix addition*, *scalar multiplication*, and *transposition* and form the basic techniques to deal with matrices.

The sum of two $m \times n$ matrices is calculated element-wise:

$$\begin{bmatrix} 1 & 3 & 1 \\ 1 & 0 & 0 \end{bmatrix} + \begin{bmatrix} 0 & 0 & 5 \\ 7 & 5 & 0 \end{bmatrix} = \begin{bmatrix} 1+0 & 3+0 & 1+5 \\ 1+7 & 0+5 & 0+0 \end{bmatrix} = \begin{bmatrix} 1 & 3 & 6 \\ 8 & 5 & 0 \end{bmatrix}$$

The scalar multiplication $c\mathbf{A}$ of a matrix $\mathbf{A}$ and a number c (also called a *scalar*) is given by multiplying every element of $\mathbf{A}$ by c:

$$2 \times \begin{bmatrix} 1 & 8 & -3 \\ 4 & -2 & 5 \end{bmatrix} = \begin{bmatrix} 2 \times 1 & 2 \times 8 & 2 \times (-3) \\ 2 \times 4 & 2 \times (-2) & 2 \times 5 \end{bmatrix} = \begin{bmatrix} 2 & 16 & -6 \\ 8 & -4 & 10 \end{bmatrix}$$

The *transpose* of an $m \times n$ matrix $\mathbf{A}$ is the $n \times m$ matrix denoted as $\mathbf{A}^\mathsf{T}$ and is formed by converting the rows of $\mathbf{A}$ into the columns of $\mathbf{A}^\mathsf{T}$, i.e., $\mathbf{A}^T_{j,i} = \mathbf{A}_{i,j}$ for $i = 1, \dots, m$ and $j = 1, \dots, n$.

$$\mathbf{A} = \begin{bmatrix} 2 & 16 & -6 \\ 8 & -4 & 10 \end{bmatrix}$$

$$\mathbf{A}^\mathsf{T} = \begin{bmatrix} 2 & 8 \\ 16 & -4 \\ -6 & 10 \end{bmatrix}$$

Familiar properties of numbers extend to the operations of matrices. For example, addition is commutative, i.e., the matrix sum does not depend on the order of the summands: $\mathbf{A} + \mathbf{B} = \mathbf{B} + \mathbf{A}$. The transpose is compatible with addition and scalar multiplication, as expressed by $(c\mathbf{A})^\mathsf{T} = c(\mathbf{A}^\mathsf{T})$ and $(\mathbf{A} + \mathbf{B})^\mathsf{T} = \mathbf{A}^\mathsf{T} + \mathbf{B}^\mathsf{T}$. Finally, $(\mathbf{A}^\mathsf{T})^\mathsf{T} = \mathbf{A}$.

Row operations are used to change matrices. There are three types of row operations: row switching (interchanging two rows of a matrix), row multiplication (multiplying all entries of a row by a non-zero constant), and, finally, row addition (adding a multiple of a row to another row). These row operations are used in a number of ways including solving linear equations and finding inverses.

Multiplication of two matrices is defined only if the number of columns of the left matrix is the same as the number of rows or the right matrix. If $\mathbf{A}$ is an $m \times n$ matrix and $\mathbf{B}$ is an $n \times p$ matrix, then their matrix product $\mathbf{AB}$ is the $m \times p$ matrix whose entries are given by the *dot product* of the corresponding row of $\mathbf{A}$ and the corresponding column of $\mathbf{B}$:

$$AB_{i,j} = A_{i,1}B_{1,j} + A_{i,2}B_{2,j} + \dots + A_{i,n}B_{n,j} = \sum_{r=1}^{n} A_{i,r}B_{r,j}$$

where $1 \le i \le m$ and $1 \le j \le p$. For example,

$$\begin{bmatrix} 1 & 0 & 2 \\ -1 & 3 & 1 \end{bmatrix} \times \begin{bmatrix} 3 & 1 \\ 2 & 1 \\ 1 & 0 \end{bmatrix} = \begin{bmatrix} 5 & 1 \\ 4 & 2 \end{bmatrix}$$

Matrix multiplication satisfies the rules $(\mathbf{AB})\mathbf{C} = \mathbf{A}(\mathbf{BC})$, i.e., *associativity*, and $(\mathbf{A} + \mathbf{B})\mathbf{C} = \mathbf{AC} + \mathbf{BC}$ as well as $\mathbf{C}(\mathbf{A} + \mathbf{B}) = \mathbf{CA} + \mathbf{CB}$, i.e., left and right *distributivity*, whenever the size of the matrices is such that the various products are defined. The product $\mathbf{AB}$ may be defined without $\mathbf{BA}$ being defined, namely if $\mathbf{A}$ and $\mathbf{B}$ are $m \times n$ and $n \times k$ matrices, respectively, and $m \ne k$. Even if both products are defined, they need not be equal, i.e., matrix multiplication in general is not *commutative*, $\mathbf{AB} \ne \mathbf{BA}$.

The *identity matrix* $\mathbf{I}_n$ of size n is the $n \times n$ matrix in which all the elements on the main diagonal are equal to 1 and all other elements are equal to 0, e.g.,

$$I_3 = \begin{bmatrix} 1 & 0 & 0 \\ 0 & 1 & 0 \\ 0 & 0 & 1 \end{bmatrix}$$

It is called an identity matrix because multiplication with it leaves a matrix unchanged: $\mathbf{MI}_n = \mathbf{I}_m\mathbf{M} = \mathbf{M}$ for any $m \times n$ matrix $\mathbf{M}$.

A particular case of matrix multiplication is tightly linked to linear equations: if $\underline{\mathbf{x}}$ designates a column vector (i.e., $n \times 1$ matrix) of n variables $x_1, x_2, \dots, x_n$, and $\mathbf{A}$ is an $m \times n$ matrix, then the matrix equation

$$\mathbf{A}\underline{\mathbf{x}} = \underline{\mathbf{b}}$$

where $\underline{\mathbf{b}}$ is some $m \times 1$ column vector is equivalent to the system of linear equations

$$A_{1,1}x_1 + A_{1,2}x_2 + \dots + A_{1,n}x_n = b_1$$

$$\dots$$

$$A_{m,1}x_1 + A_{m,2}x_2 + \dots + A_{m,n}x_n = b_m$$

This way, matrices can be used to compactly write and deal with multiple linear equations, i.e., systems of linear equations.

Index

Printed and bound by CPI Group (UK) Ltd, Croydon, CR0 4YY

18/10/2024

01776264-0002